T0212235

COMPUTER ARCHITECTURE TECHNIQUES FOR POWER-EFFICIENCY

Synthesis Lectures on Computer Architecture

Editor

Mark D. Hill, *University of Wisconsin, Madison*

Synthesis Lectures on Computer Architecture publishes 50 to 150 page publications on topics pertaining to the science and art of designing, analyzing, selecting and interconnecting hardware components to create computers that meet functional, performance and cost goals.

Computer Architecture Techniques for Power-Efficiency
Stefanos Kaxiras and Margaret Martonosi
2008

Chip Mutiprocessor Architecture: Techniques to Improve Throughput and Latency
Kunle Olukotun, Lance Hammond, James Laudon
2007

Transactional Memory
James R. Larus, Ravi Rajwar
2007

Quantum Computing for Computer Architects
Tzvetan S. Metodi, Frederic T. Chong
2006

Computer Architecture Techniques for Power-Efficiency

Stefanos Kaxiras and Margaret Martonosi

ISBN: 978-3-031-00593-0   paper
ISBN: 978-3-031-01721-6   ebook

DOI: 10.1007/978-3-031-01721-6

A Publication in the Springer series
SYNTHESIS LECTURES ON COMPUTER ARCHITECTURE #4

Lecture #4
Series Editor: Mark D. Hill, University of Wisconsin, Madison

Library of Congress Cataloging-in-Publication Data

Series ISSN: 1935-3235   print
Series ISSN: 1935-3243   electronic

COMPUTER ARCHITECTURE TECHNIQUES FOR POWER-EFFICIENCY

Stefanos Kaxiras
University of Patras, Greece
Kaxiras@ece.upatras.gr

Margaret Martonosi
Princeton University
mrm@princeton.edu

SYNTHESIS LECTURES ON COMPUTER ARCHITECTURE #4

ABSTRACT

In the last few years, power dissipation has become an important design constraint, on par with performance, in the design of new computer systems. Whereas in the past, the primary job of the computer architect was to translate improvements in operating frequency and transistor count into performance, now power efficiency must be taken into account at every step of the design process.

While for some time, architects have been successful in delivering 40% to 50% annual improvement in processor performance, costs that were previously brushed aside eventually caught up. The most critical of these costs is the inexorable increase in power dissipation and power density in processors. Power dissipation issues have catalyzed new topic areas in computer architecture, resulting in a substantial body of work on more power-efficient architectures. Power dissipation coupled with diminishing performance gains, was also the main cause for the switch from single-core to multi-core architectures and a slowdown in frequency increase.

This book aims to document some of the most important *architectural* techniques that were invented, proposed, and applied to reduce both dynamic power and static power dissipation in processors and memory hierarchies. A significant number of techniques have been proposed for a wide range of situations and this book synthesizes those techniques by focusing on their common characteristics.

KEYWORDS

Computer power consumption, computer energy consumption, low power computer design, computer power efficiency, dynamic power, static power, leakage power, dynamic voltage/ frequency scaling, computer architecture, computer hardware.

Contents

Acknowledgements

Stefanos:

I would like to thank my co-author Margaret Martonosi for our great collaboration. She has been an inspiration for me since I met her. My thanks to Mark Hill and Michael Morgan for seeing this project from conception to publication.

My wife, Angeliki, and our children Jason and Eugenia encouraged me throughout the writing of this book with their love and laughter; and helped me move along with their tireless prompts to "finish this book at last." I owe them my love and thanks.

Many people helped by reading drafts, offering suggestions, and having detailed discussions with us on book topics. I would especially like to thank Georgios Keramidas, Chronis Xekalakis, Vasileios Kontorinis, Daniele Ludovici, Ioannis Sourdis, and Christos Strydis, who have spent many hours reading drafts of this book. I am indebted to Georgios Keramidas for his help with the material in Section 4.12.

My European colleagues in the HiPEAC Network of Excellence and the EU SARC Integrated Project were especially patient with me during the past year. I thank all of them for their understanding. My thanks to Erik Hagersten for his support; a portion of the book was written in Sweden at a time when I was working for Erik's company and teaching at Uppsala University.

Finally, I would like to honor the memory of two dear friends and colleagues, Stamatis Vassiliadis and David V. James who, sadly, passed away in the last 12 months.

Margaret:

First and foremost, I thank my colleague Stefanos Kaxiras; without his wisdom, enthusiasm, and leadership, this book would not have happened. Likewise, I thank Mark Hill and Michael Morgan for extending us the opportunity to work on this project. Thanks also to the reviewers who offered us excellent feedback on how to make improvements to the presentation after the first draft.

My graduate students over the years have, through their outstanding research, continually taught me more and more about computer architecture in general and about power-efficiency in particular. This book comprises—directly and indirectly—many of their contributions to the field. Finally, I thank my husband, Kevin Burkman. In addition to being my best friend, he has also been my tireless coach on the value of non-procrastinatory behavior. As I write this (unfortunately at the last minute!), I thank him for these two roles (and the many others) he plays in my life.

CHAPTER 1

Introduction

In the 1980s and early 1990s, computer architecture experienced a phase in which quantitative performance evaluations became the predominant driver for determining how to build effective, cutting-edge microprocessors and computer systems. While cost, area, and other metrics remained important as limits to abide by, performance was during this period the main goal for which designers optimized.

In the mid- to late- 1990s, power began to be an area of concern for architects. It became clearer that as Moore's Law scaling succeeded in reducing the feature sizes of semiconductor devices, their areal power density and high processor clock rates would result in microprocessor designs that were difficult or impossible to cool.

Interestingly, the CMOS power issues that our field currently faces are not the first instance of power constraints affecting computer systems design. This chapter will first give a history of power issues in computer systems, then lay out the trends and issues that lie ahead.

1.1 BRIEF HISTORY OF THE "POWER PROBLEM"

Power and cooling concerns are not strictly a twenty-first century issue for computing. In reality, prior computing eras have also faced power and thermal challenges. For example, the ENIAC machine built in 1947 dissipated 174 kW (233 horsepower) [29, 59]! A March 1949 article in *Popular Mechanics* extrapolated from ENIAC's 18 000 vacuum tubes toward a day in the future where computers might need only 1000 vacuum tubes and therefore presumably drop toward only 10 kW of power [92].

Figure 1.1 illustrates a time history of power trends for selected computers from 1950 to the present. The y-axis of this plot shows areal power density, that is watts per unit surface area. While very early (vacuum tube) computers were extremely high in power consumption, their large areas kept power density relatively low. Furthermore, the switch from high-power vacuum tubes to relatively lower-power bipolar transistors kept power dissipation manageable for some time. For example, the Intel 4004 microprocessor had, in 1971, similar compute capability as the multi-ton, room-sized ENIAC computer, and yet dissipated only a handful of watts.

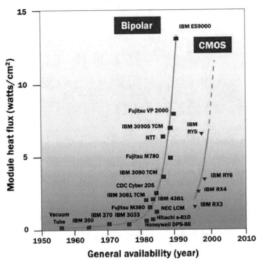

FIGURE 1.1: A history of areal power density trends across decades of computing technologies. Reproduced from [54]. Copyright 1999 IEEE.

During the 1980s, the widespread use of bipolar semiconductor technologies and increasingly dense transistor integration led to rapid rises in power dissipation and density. Challenges in cooling and power delivery were not uncommon. For example, in this time period, IBM and Cray produced servers and supercomputers that required liquid cooling in order to maintain their very high performance targets [144, 160]. As another example, the BIPS research processor [121, 120] was notable in the early 1990s both due to its high performance and very high clock rate: 300 MHz, as well as due to its record-setting power dissipation: 115 W.

The shift from bipolar technologies to CMOS technologies brought temporary relief from power challenges, as shown in the timeline in the early 1990s. CMOS technology had been in use before the 1990s, and had appealing power behavior in the sense that it primarily dissipated power only at switching transitions. The complementary gate structure meant that early gates drew little or no current between transition points, because in a stable state, the gate has no clear path to ground. (At this point, CMOS technology scaling had not created significant leakage paths yet.)

During the transition period of the 1980s, many viewed CMOS as too slow for widespread use in the high-performance microprocessor arena. Though researchers in semiconductor device technologies explored various alternatives (such as hybrids of bipolar and CMOS), the power challenges with bipolar, however, became too great to ignore. These challenges drove the switchover to CMOS, and along the way, technology improvements brought significant gains to CMOS performance.

1.2   CMOS POWER CONSUMPTION: A QUICK PRIMER

The remainder of this book focuses primarily on the issues facing computer architects in managing and optimizing CMOS power dissipation. We offer a brief primer here on these issues, with more detail given in Chapter 2.

CMOS power consumption can be divided into several categories: dynamic power, leakage power, glitching power, and others. We discuss these in the subsections that follow.

1.2.1   Dynamic Power

The dominant power category for many years has been dynamic power, which is given by the proportionality: $P = $ (proportional to) CV^2Af. Here, C is the load capacitance, V is the supply voltage, A is the activity factor and f is the operating frequency. Each of these is described in greater detail below.

Capacitance (C): To first order (i.e., the architect's point of view rather than the more detailed view of a circuit designer), aggregate load capacitance largely depends on the wire lengths of on-chip structures. Architects can influence this metric in several ways. As one example, building four smaller processor cores on-chip, rather than one large monolithic processor, is likely to reduce average wire lengths considerably, since most wires will interconnect units within a single core. Likewise, smaller cache memories or independent banks of cache can also reduce wire lengths since many address and data lines will only need to span across each bank array individually.

Supply voltage (V): For decades, supply voltage (V or V_{dd}) has dropped steadily with each technology generation [113]. Because of its direct quadratic influence on dynamic power, this has amazing leverage on power-aware design.

Activity factor (A): The activity factor is a fraction between 0 and 1 that refers to how often wires actually transition from 0 to 1 or 1 to 0. While the clock signal obviously switches at its full frequency, most other wires in the design have activity factors below 1. Strategies such as clock gating are used to save energy by reducing activity factors during a hardware unit's idle periods. In particular, the clock gating technique ANDs a unit's clock signal with a control signal. When the control signal is 1, the unit will be clocked as expected. If the unit is known to be unneeded for a cycle or more, the control signal can be set to 0, in which case the unit will not be clocked; this can reduce the switching activity within it.

Clock frequency (f): The clock frequency has a fundamental and far-reaching impact on power dissipation. Not only does clock frequency directly influence power dissipation, but it also indirectly shapes power by its effect on supply voltage. Typically, maintaining higher clock frequencies may require (in part) maintaining a higher supply voltage. Thus, the combined V^2f portion of the dynamic power equation has a cubic impact on power dissipation. Strategies such as dynamic voltage and frequency scaling (DVFS) recognize periods when lower microprocessor performance is acceptable (e.g., in memory-bound or latency-tolerant regions of code) and reduce (V, f) accordingly.

1.2.2   Leakage

While dynamic power dissipation represented the predominant factor in CMOS power consumption for many years, leakage energy has been increasingly prominent in recent technologies. Representing roughly 20% or more of power dissipation in current designs, its proportion is expected to increase in the future [32, 113]. Leakage energy can come from several sources, including gate leakage and sub-threshold leakage. Gate leakage is increasing in importance and will be discussed in Chapter 5. Here, we briefly introduce only the concepts behind sub-threshold leakage because they are fundamental to this chapter's trends discussion.

Sub-threshold leakage power represents the power dissipated by a transistor whose gate is intended to be off. While our idealized view of transistors is that they operate as switches, the reality is that the relationship between current and voltage (the so-called *IV* curve depicted in Figure 1.2) is analog and shows a non-zero amount of current even for voltages lower than the threshold voltage (V_{th}) at which the transistor is viewed as switching "on." This modest current for V_{dd} less than V_{th} is referred to as the sub-threshold current. The power dissipation resulting from this current is referred to as the sub-threshold leakage power, because the transistor appears to leak charge to ground. Sub-threshold leakage power is given by the following *simplified* equation:

$$P = V\left(k\mathrm{e}^{-q V_{\text{th}}/(a k_{\text{a}} T)}\right).$$

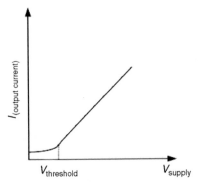

FIGURE 1.2: Example of an "*IV*" curve for a semiconductor diode. Although we informally treat semiconductors as switches, their non-ideal analog behavior leads to leakage currents and other effects.

In this equation, V refers to the supply voltage, while V_{th} refers to the threshold voltage. The exponential link between leakage power and threshold voltage is immediately obvious.[1] Lowering the threshold voltage brings a tremendous increase in leakage power. Unfortunately, lowering the threshold voltage is what we have to do to maintain the switching speed in the face of lower supply voltages. Temperature, T, is also an important factor in the equation: leakage power depends exponentially on temperature. The remaining parameters, q, a, and k_a, summarize logic design and fabrication characteristics. The exponential dependence of leakage on temperature, and the interplay between leakage and dynamic energy will be discussed in more detail in Chapter 2.

1.2.3 Other Forms of CMOS Power Dissipation

While dynamic and leakage power dominate the landscape, other forms of power dissipation do exist. For example, short circuit or "glitching" power refers to the power dissipated during the brief transitional period when both the n and p transistors of a CMOS gate are "on," thus forming a short-circuit path from power to ground. This is distinguished from dynamic power because dynamic power typically refers to power dissipated due to discharging charged capacitors; it would be dissipated even if transitions occurred instantaneously. In contrast, glitching power refers to transitional power that occurs because of non-ideal transition times.

1.3 POWER-AWARE COMPUTING TODAY

From the early 1990s to today, power consumption has transitioned into a primary design constraint for nearly all computer systems. In mobile and embedded computing, the connection from power consumption to battery lifetime has made the motivation for power-aware computing very clear. Here, it is typically low energy that is stressed, although obviously power/performance is also important for some embedded systems with high computational requirements.

In desktop systems, the key constraint has become thermal issues. Excessive power consumption is one of the prevailing reasons for the abrupt halt of clock frequency increases. Currently, high-performance processor clocks have hit a "power wall" and are pegged at or below 4 GHz. This is contrary to 2001 ITRS projections which predicted clocks in excess of 6 GHz by roughly 2006. Power consumption is also one important factor driving the adoption of chip multiprocessors (CMPs) since they allow high-throughput computing to be performed within cost-effective power and thermal envelopes.

---

[1]What is not shown in this simplified equation is the—also exponential—dependence of leakage power to the supply voltage. This is discussed in Chapter 5.

In servers and data centers, energy and thermal issues are important on an aggregate scale. Recent internet data centers are estimated to draw in 50 MW of aggregate power or more [175]. On the aggregate scale, reducing processor power can have high leverage; a rough rule of thumb is that 1 W of power saved in the processor translates into an additional watt saved in power supply efficiency, and another 1 W saved in cooling requirement reductions. A recent report by the Boyd Co. indicates that within the United States, even the least expensive possible data center sites will result in annual operating costs of roughly $10M per year for a data center of 75 employees and electricity plays an increasingly major role in data center siting costs [71]. For example, HSBC's decision to build a large data center near Buffalo, NY is said to have been strongly influenced by a New York State incentive package including 11 MW of cheap hydroelectric power. Likewise, Google, Microsoft, and Yahoo are all said to be building large data centers along the Columbia River in Washington and Oregon for proximity to inexpensive electricity [203].

Among researchers, circuits and VLSI specialists focused on the power problem much earlier than architects. This comes as no surprise since people in circuits and VLSI came into contact with chip power budgets well before architects. They also have more direct tools for analyzing power issues (late in the design timeline) and direct circuit techniques to address some of them. While architects addressed problems later than the "lower" hardware fields, their advantage is in leverage. Addressing power issues early and holistically in the design process has the potential for better and more adaptable power-performance tradeoffs.

By the late 1990s, power was universally recognized by architects and chip developers as a first-class constraint in computer systems design. Today, power cannot be ignored in any new microarchitectural proposal. At the very least, a microarchitectural idea that promises to increase performance must justify not only its cost in chip area but also its cost in power. Thus, much of the research described in this book was proposed in the last ten years.

1.4   THIS BOOK

The target readers of this book are engineers or researchers who are fairly fluent in computer architecture concepts, but who want to build their understanding of how power-aware design influences architectures. We envision a computer architecture graduate student or advanced undergraduate, as well as industry engineers. We write this without assuming detailed knowledge of transistor or circuits details, beyond the basics of CMOS gate structures.

In addition to offering background information on how and why power trends arise, we also see the book as a compendium of basic strategies in power-aware design. While no book of this length could enumerate all possible power-saving techniques, we try to include the most fundamental ones known to the field as we write this in the Summer of 2007.

The structure of the book is as follows. Chapter 2 offers deeper background information on power dissipation and describes the primary strategies for modeling, simulating, and measuring power and related metrics. The rationale for this chapter is that one cannot optimize power for a particular system if one lacks a clear view of that system's power behavior. Therefore, we view Chapter 2 as offering resources on experimental and measurement infrastructure that every architect can tailor to the needs of their research or design.

Chapters 3 and 4 cover aspects of reducing dynamic power in CMOS computer systems. We have chosen to arrange this material in terms of the basic equation for CMOS dynamic power consumption: CV^2Af. Thus, Chapter 3 covers strategies for managing power via voltage (V) and clock frequency (f). This includes both dynamic voltage and frequency scaling, as well as other possibilities.

Chapter 4 focuses on the activity factor (A) and capacitance (C). Activity factor is so intrinsically tied to how architectural units are organized and used that it represents much of the most straightforward power optimizations available to a computer architect. Other important methods of reducing power consumption manage the capacitance (C) factor in designs. At the most qualitative but intuitive level, shorter wires have lower capacitance; thus, microarchitectures with simple local structures are likely to result in improved dynamic power behavior. Such insights have driven many power-aware designs, such as memory banking optimizations and even the overwhelming current trend towards chip multiprocessors (CMPs). We have grouped the discussion of "A" and "C" together because they are often linked in design strategies. For example, one can reduce the activity factor on buses or in arithmetic units by segmenting long wires into individually controllable modules; this affects both C and A.

While dynamic power is important and represents much of the prior work in power-aware architecture, there is no avoiding the fact that leakage energy has emerged as an equal or greater challenge for computer architects today. Thus, in Chapter 5 we present a discussion of techniques for lower static or leakage power in current and future computer systems. Finally, Chapter 6 offers our conclusions.

CHAPTER 2

Modeling, Simulation, and Measurement

As the power problem has become prominent for computer architects, many ideas have been proposed for managing power and energy issues through architectural techniques. In order to compare these many ideas, quantitative techniques for architecture-level power modeling have become very important. Thus, in this chapter, we discuss some of the key issues and techniques for the field of architecture-level power modeling. This sets the foundation for later chapters in which power-efficient ideas are discussed and compared.

2.1   METRICS

The metric of interest in power studies varies depending on the goals of the work and the type of platform being studied. This subsection offers an overview of possible metrics, and discusses the best practices regarding when to use them.

Energy: Energy, in joules, is often considered the most fundamental of the possible metrics, and is of wide interest particularly in mobile platforms where energy usage relates closely to battery lifetime. Even in non-mobile platforms, energy can be of significant importance. For data centers and other "utility computing" scenarios, energy consumption ranks as one of the leading operating costs and thus reducing energy usage is crucial [71, 203]

Power: Power is the rate of energy dissipation. The unit of power is watts (W), which is joules per second. Power's time component makes it the meaningful metric for understanding current delivery and voltage regulation on-chip. Another related metric, areal power density, is power per unit area. This metric is useful for thermal studies; 200 W spread over many square centimeters may be quite easy to cool, while 200 W dissipated in the relatively small ($\sim$4 cm^2) areas of today's microprocessor dies becomes challenging or impossible to cool in a cost-effective manner.

Energy-per-instruction: In some situations, however, focusing solely on energy is not enough. For example, reducing energy at the expense of lower performance may often not be acceptable. Thus, metrics combining energy and performance have been proposed. For example, energy-per-instruction (EPI) is sometimes used as a method of comparing energy optimizations, particularly those that focus on general microarchitectural traits, rather than on the runtime of a particular application.

Energy-delay product: While low power often used to be viewed as synonymous with lower performance, that is no longer the case. In many cases, application runtime is of significant relevance even in energy- or power-constrained environments. With the dual goals of low energy and fast runtimes in mind, energy-delay product (EDP) was proposed as a useful metric [85]. EDP offers equal "weight" to either energy or performance degradation. If either energy or delay increase, the EDP will increase. Thus, lower EDP values are desirable. When comparing scenarios that do not alter the instruction count or mix, EDP is roughly equivalent to the reciprocal of MIPS^2/Watt. Note the derivation below:

$$Delay = runtime$$
$$Energy = Watts * runtime$$
$$EDP = Watts * runtime * runtime.$$
$$runtime = Instruction\ Count\ /\ MIPS$$
$$EDP = Watts * (ICount\ /\ MIPS)^2$$
$$EDP = ICount^2 * 1/(MIPS^2/Watt).$$

Unlike EPI, EDP's inclusion of runtime means that this is a metric that *improves* for approaches that either hold energy constant but execute the same instruction mix *faster*, or hold performance constant but execute at a lower energy, or some combination of the two.

Energy-delay-squared and beyond: Following on the original EDP proposal, other work has suggested alternative metrics, such as energy-delay-squared product (ED^{2}P) or energy-delay-cubed product (ED^{3}P) [211, 251]. These alternatives correspond to MIPS^{3}per Watt or MIPS^{4}per Watt. At a qualitative level, one can view these metrics as applying to the high-performance arena where performance improvements may matter more than energy savings.

Delving deeper into these metrics, one can argue that ED^{2}P makes the most sense when considering fixed microarchitectures, but accounting for voltage scaling as a possible energy management technique. In particular, consider the following rough trends: power is proportional to CV^2f, which for a fixed microarchitecture and design is proportional to V^3. Performance, on the other hand, is roughly proportional to frequency. Since frequency varies roughly linearly with voltage in the 1–3 V range, this means that performance is also roughly proportional to voltage. As a result, when processors use voltage scaling as a primary power-performance trade-off, metrics considering (perf)^3 / power are the fair way to compare energy efficiencies. This, in fact, is ED^{2}P or MIPS^3/Watt.

The broader question of how to weigh energy and performance is often answered specifically in regards to particular designs, or even to particular modules and decisions within a

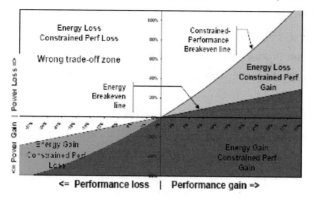

FIGURE 2.1: Performance/power tradeoff zones. Reproduced from [84]. Copyright 2003 Intel.

design. In discussing design decisions in Intel's Pentium M processor, Gochman et al. used a particularly compelling graphic to convey the different options [84]. Figure 2.1 reproduces it here.

The figure depicts performance gains (or loss) on the *x*-axis, and power improvements or degradations on the *y*-axis. Clearly, a technique that degrades both power and performance is of little appeal; these lie in the upper-left quadrant of the graph. The other three quadrants contain regions that represent possible power-performance options under different constraints. The magenta region represents the realm in which both performance and energy are improved. The orange region represents cases where performance improves (e.g. subject to a thermal constraint) despite overall energy loss, and the green region shows where energy improves despite a loss in performance.

Other metrics and issues: The remainder of this chapter and book focuses on modeling and research ideas primarily in the context of the above metrics. There are other metrics and design goals that are also relevant, although space constraints preclude us from discussing them in detail. In particular, we briefly discuss thermal modeling because of its relevance to leakage energy modeling, but we cannot discuss thermal-aware design techniques in deep detail.

2.2   MODELING BASICS

This section gives an overview of modeling techniques for each of the primary power-related issues: dynamic power, leakage power, and thermal behavior. Because issues of modeling and simulation are particularly intertwined for leakage power and thermal behavior, we cover simulators and simulation in these sections as well. Dynamic power simulation, however, is a more mature and well-trodden area; as such, we cover it in Section 2.3 which follows.

2.2.1   Dynamic-power Models

The most widely used and well-understood power models among architects are those that focus on dynamic power consumption. Because dynamic power greatly exceeded leakage power until recently, it is reasonable that early architects focused their attention on how to measure and model its effects.

Drawing from the familiar CV^2Af equation previously presented, dynamic power models typically focus on characterizing these terms. We start by considering a scenario in which V and f are viewed as fixed, focusing mainly on C and A. We then move to consider other approaches in which power predictions across technology generations (and therefore spanning values of V and f) are attempted.

At a high level, dynamic-power models can be divided into analytical and empirical techniques. Analytical techniques seek to express power behavior in terms of equations parameterized by module size or other characteristics. Empirical techniques, in contrast, have focused on predicting the behavior of one possible chip design by appropriately scaling per-module power behaviors observed for some other *measured* chip design. We focus on analytic models here, but discuss empirical approaches when we touch on dynamic power simulation in Section 2.3.

Early work in the architecture-level power modeling mainly focused on caches [20, 119, 123, 213]. This was in part due to the fact that caches represented up to 40% of the power budget for some low-power embedded microprocessors [169]. In addition, caches are regular structures which are easier to model. Both C and A can be fairly readily expressed using parameterizations of cache size and organization. Thus, it makes sense that attention would be focused here earlier.

Both capacitance and activity factor are expressions where the architect has some high-level understanding and control, even though the ultimate details are dependent on the particulars of the circuit design chosen.

The activity factor is related both to the application program being executed (both its data patterns and control) and to some circuit design choices. For example, for circuits that pre-charge and discharge on every cycle (i.e., double-ended array bitlines) an A of 1 is used. For wires that represent data buses, the activity factor can be chosen based on knowledge of the 1/0 statistics in the data set being studied. In addition, clock gating (techniques that "gate" the clock control signal to prevent its toggling within a particular sub-unit) reduce a sub-unit's activity factor during idle periods.

Estimating Capacitance: Like activity factor, capacitance depends in part on circuit design choices. Even in relatively regular array structures, the aspect ratio, number of wire routing layers, or other layout choices can influence capacitance. Nonetheless, with modest amounts of knowledge about circuit design style, usefully accurate architecture-level power models are

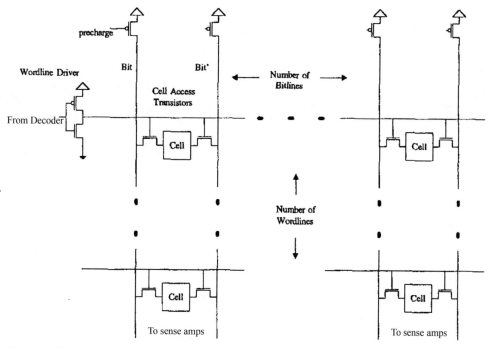

FIGURE 2.2: The regularity of array structures such as simple caches allows them to be modeled using a parameterized approach based on the number of bitlines and wordlines. (Figure taken from [38]).

possible. Figure 2.2 illustrates a possible model layout for array structures, and Table 2.1 shows a set of parameterized capacitance estimators for a register file [38].

2.2.2 Leakage Models

Leakage energy did not capture the attention of architects until somewhat later than dynamic energy. Its importance, however, has grown considerably over the past five years. In part, this importance stems directly from the fact that leakage energy now represents 20–40% of the power budget of microprocessors in current and near-future fabrication technologies [32, 119]. In addition, leakage energy plays a role in a harmful feedback loop with thermal issues, as we will discuss below.

The original and still primary model of leakage behavior used by computer architects centers around the following equation:

$$I_{\text{leak}} = \mu_0 C_{\text{OX}} \frac{W}{L} e^{a+b^*(V_{\text{dd}}-V_{\text{dd0}})} v_{\text{t}}^2 \left(1 - e^{\frac{-V_{\text{dd}}}{v_{\text{t}}}}\right) \exp\left(\frac{-|V_{\text{th0}}| - V_{\text{off}}}{n \cdot v_{\text{t}}}\right).$$

In this equation, some of the parameters are clearly related to fabrication technology and can largely be viewed by computer architects as constants. Thus, Butts and Sohi [41] created a

TABLE 2.1: This Table Summarize Parameterized Equations That Express the Capacitance for Different Key Nodes of the Cache Array in Figure 2.2. C_{diff} C_{gate}, and C_{metal} refer to the capacitance of diffusion, gate and metal regions respectively. Adapted from [38].

Node	Capacitance Equation
Regfile Wordline Capacitance =	$C_{diff}(WordLineDriver) + C_{gate}(CellAccess)$ $* NumBitlines + C_{metal} * WordLineLength$
Regfile Bitline Capacitance =	$C_{diff}(PreCharge) + C_{diff}(CellAccess) * NumWdlines$ $+ C_{metal} * BLLength$
CAM Tagline Capacitance =	$C_{gate}(CompareEn) * NumberTags$ $+ C_{diff}(CompareDriver) + C_{metal} * TLLength$
CAM Matchline Capacitance =	$2 * C_{diff}(CompareEn) * TagSize$ $+ C_{diff}(MatchPreCharge)$ $+ C_{diff}(MatchOR) + C_{metal} * MLLength$
ResultBus Capacitance =	$0.5 * C_{metal} * (NumALU * ALUHeight)$ $+ C_{metal} * (RegfileHeight)$

model in which constants are subsumed into a clearer form, expressing I_{leak} as

$$= N \cdot K_{design} \cdot k_{tech} \cdot 10^{-V_T/S_T}.$$

From this, P_{leak} can be expressed as

$$P_{static} = V_{CC} \cdot N \cdot k_{design} \cdot I_{leak}.$$

The key insight here is that many detailed aspects of the circuit design choices can be abstracted into the k_{design} factor seen in these equations. The k_{design} parameter in some sense represents the degree of "stacking" seen by transistors in different types of circuit designs (e.g., array structures versus static logic, etc.). The HotLeakage simulation framework [247] builds on Butts/Sohi analytics to provide a simulation package for leakage energy.

To see how Butts and Sohi arrived at the simplified formulas, we start with the Berkeley Predictive Model (BSIM3V3.2) formula that gives subthreshold leakage current as

$$I_{Dsub} = I_{s0} \cdot \left(1 - e^{\frac{-V_{ds}}{v_t}}\right) \cdot e^{\frac{V_{gs}-V_T-V_{off}}{n \cdot v_t}}$$

In this equation V_{ds} is the voltage across the drain and the source and V_{gs}, the voltage across the gate and the source terminal. V_{off} is an empirically determined model parameter and v_t is a physical parameter proportional to temperature—the exponential dependence to temperature is immediately obvious. The term n encapsulates various device parameters. The term I_{s0} depends on transistor geometry and can be written as $I_{s0} \times W/L$. The Butts and Sohi model examines and simplifies the above equation for a single device in its normal "off" state, where $V_{ds} = V_{CC}$ and $V_{gs} = 0$. This makes the factor

$$\left(1 - e^{\frac{-V_{ds}}{v_t}}\right)$$

approximately 1 since $V_{ds} = V_{dd} \gg V_T$. By grouping more terms together, Butts and Sohi simplify the formula to

$$I_{Dsub} = \left(\frac{W}{L}\right) \times k_{Tech} \times 10^{\left(\frac{-V_T}{s_t}\right)}.$$

The simplified formula exposes only the relationship of leakage to transistor geometry, to threshold voltage (V_T), and indirectly to temperature via the s_t term. Accounting for the many (N) similarly sized transistors with similar characteristics in larger structures further simplifies the formulas by encapsulating the various (W/L) terms of each transistor into a new parameter k_{design}. The formula for the static *power* consumption for the N transistors can then be expressed as:

$$P_{leakage} = V_{dd} \times N \times k_{design} \times k_{Tech} \times 10^{\left(\frac{-V_T}{s_t}\right)}.$$

The parameters k_{design} and k_{Tech} can be looked up from tables such as Table 2.2 for k_{design}. The number of devices, and the supply and threshold voltages are the only concerns of the architect for estimating the leakage power of a design.

2.2.3   Thermal models

A remaining important piece of the modeling landscape is the ability to model thermal behavior. A cyclic relationship exists between power and thermal modeling. On one hand, thermal behavior depends on power dissipation and density, since temperature is essentially a function of how much power is dissipated in a region versus how that region is cooled. On the other hand, power also depends on temperature. This is particularly true due to the exponential dependence of leakage energy on temperature as indicated in the equations in Section 2.2.2.

In addition to this cyclic dependence of power and temperature, another wrinkle in thermal modeling concerns the timescale of interest for the model. For example, in terms of provisioning the cooling capacity of a chip, a system, or a data center, long-term "steady-state" temperature may be the metric of interest. In terms of either designing a microprocessor or

TABLE 2.2: K_{design} parameters for Typical Circuits. Adapted from [41].

Circuit	N	K_{design}	Notes
D flip flops	22/bit	1.4	Edge-triggered FF
D latch	10/bit	2.0	Transparent latch
2-input Mux	2/bit/input	1.9	+1.2/input over 2
6T RAM cell	6/bit	1.2	1 RW port
CAM cell	13/bit	1.7	1 RW, 1 CAM port
Static logic	2/gate input	11	Depends on speed, load, ±3

crafting OS-level management techniques, however, more localized heating information is almost always necessary.

There are analogies between heat transfer and electrical phenomena upon which we can build thermal models. Power dissipation results in heat, and this heat flows through regions based on their thermal resistance (R). The amount of heat flow can be analogized to current (I), and the heat difference between two regions on a chip is analogous to voltage (V). Because there are time dependences in both the power dissipation and in its relationship to heat flow and thermal impedance, a capacitance (C) is also modeled. Thus, time-dependent RC models remain the best way to model localized thermal behavior on chip.

In some early work, TEMPEST modeled temperature based on power dissipation and density values, but did so only for the chip as a whole, not for individual regions or architectural units [65].

Perhaps the most important work on architecture-level thermal modeling thus far been the HotSpot approach developed by Skadron et al. [206]. They propose and validate a compact RC model for localized heating in high-end microprocessors. This model considers both the lateral relationships between units on chip, as well as the vertical heating/cooling relationships between the active portion of the silicon die and the attached heat spreader and heat sink layers that seek to even out temperature and draw heat away from the active silicon.

There are several steps in producing a HotSpot thermal model. First, from a chip floorplan and other technical details, one needs to deduce the R and C values that comprise the heat flow network for the RC model. In general, thermal resistance and capacitance relate to the material's thickness and the surface area of heat flow. In addition, a thermal constant k represents the material characteristics that influence heat conductivity, and another constant c represents the material's heat capacity:

$$R = t/kA,$$
$$C = ctA.$$

Such R and C values are computed both based on the areas (A) of different microarchitectural units, as well as on the areas, thickness (t), and materials of heat sinks and heat spreaders. Thus, these RC networks can be composed automatically from parameters that give the area and makeup of the microarchitecture and supporting heatsinks.

Once the RC network has been formed, it is translated into a set of differential equations, and is solved numerically using a fourth-order Runge–Kutta method. Because temperature varies slowly relative to processor speeds, the numerical solver need not be invoked on every simulated cycle. Instead, typical approaches involve a solver roughly every 10 000 processor cycles in order to track thermal trends on timescales of tens to hundreds of microseconds.

2.3   POWER SIMULATION

While the circuit design and design automation research communities researched CMOS power issues earlier, architects first began studying power issues for CMOS designs in earnest in the early to mid 1990s [88]. At first, architecture researchers interested in studying power optimizations reported their quantitative results in terms of "proxy" metrics. For example, Grunwald et al. studied power savings garnered by using confidence estimation to limit branch speculation in cases where the branch was not very likely to succeed [88]. In this work, they reported their power savings in terms of how many fewer mis-speculated instructions were executed when confidence estimators are used.

While metrics like "mis-speculation reduction" can be useful and intuitive proxies for reporting some results, their drawback is that they do not offer a common currency by which to compare the power benefits of multiple distinct power-saving opportunities. Furthermore, they do not extend naturally to studies of thermal issues and other power-related problems.

For these reasons, architects in the late 1990s began working on architecture-level power models that aim to directly estimate power and energy, just as cycle-level architecture simulators aim to directly estimate performance.

Memory system simulation: An early example of power simulation came from the Cacti tool. The Cacti tool was developed to study memory hierarchies in detail. While Cacti 1.0 [224] primarily provided estimates of area and latency for parameterized cache designs, Cacti 2.0 added in a dynamic power model. Because caches are almost always built as very regular array structures of SRAM cells with supporting circuitry, it is natural that parameterized models for their behavior would include power estimates first. Subsequent versions of Cacti have refined the dynamic power models and included leakage power models as well [217].

Whole-processor power simulators: While local power models for individual processor structures are a useful first step, processor design still required that architects be able to make high-level, whole-processor analyses of power trade-offs, and to do so early enough in the design pipeline that useful adjustments could be chosen. Thus, in 2000, two whole-processor power simulation tools were discussed. SimplePower was introduced as a means of doing detailed "whole processor" analyses of dynamic power. It focused on in-order five-stage pipelines, with detailed models of integer ALU power as well as other regions of the chip.

Also introduced in 2000, the Wattch tool, like SimplePower, sought to provide detailed whole-processor data on dynamic power [38]. Because it was built as an additional software module to be used with the widely used SimpleScalar tool [40], Wattch has seen wide use among architects. Wattch draws its cache modeling from Cacti, while providing parameterized activity-based estimates for other units as well. Wattch simulates an out-of-order super-scalar pipeline.

Both SimplePower and Wattch are examples of simulators based on analytic power modeling techniques. In contrast, the IBM PowerTimer tool represents a microprocessor power simulator based on empirical techniques [36]. Namely, PowerTimer estimates the power consumption of a particular architectural module by using the *measured* power consumption of the corresponding module in an existing reference microprocessor, and scaling it appropriate to the size and design changes. For example, if the modeled design is identical to the reference design except for a larger first-level cache, then all of its per-module power estimates would be drawn from the reference chip's measurements. For accesses to the first-level cache, the power to be "charged" would be calculated by the reference chip's cache power scaled by the expected power scaling factor. This scaling factor would most simply be based on capacitance changes, but might also include other more sophisticated effects based on cache design and layout.

Empirical power models tend to be most used in industry settings, because these architects are able to access detailed power measurements from previous reference designs. While they are quite useful for power projections into future variants of a design, they are more difficult for larger design changes where the per-module proportional scaling cannot easily be applied.

2.4 MEASUREMENT

While simulation is appealing for early-stage design evaluations, it is difficult or impossible for simulators to be deeply detailed and have sufficient speed for thorough parameter explorations. Thus, for some studies, it becomes appealing to measure power/thermal metrics directly, rather than simulating them.

Measuring total system power is, in fact, not particularly challenging. Data acquisition systems or even simple ammeters can be used to collect such aggregate numbers. Likewise, coarse-grained thermal measurements are also possible, by using software to read the on-die temperature sensor that is sometimes made available to software [196].

Challenges do exist, however, in using real-system measurements to glean deeper information about system behavior. For example, consider the deceptively simple-looking task of apportioning dynamic power into components that correspond to different hardware units on the die. While off-chip ammeters can be used to deduce how much total power the chip dissipates, there is no straightforward approach for users to determine a unit-by-unit power breakdown. To respond to this challenge, Section 2.4.1 discusses a method in which hardware performance counters are used as proxies for on-chip activity factors, in order to estimate component-level power dissipation. The section then extends on this technique to show how it can be used for thermal estimates as well.

2.4.1 Performance-Counter-based Power and Thermal Estimates

In essence, the simulation-based power estimators discussed earlier in this chapter use various approaches to estimate capacitance, and then use cycle-level simulators to estimate the "activity factors" indicating how often wires switch from zero to one or vice versa. Such approaches are appealing because they allow power estimation before a system is built, and because they allow one to explore parameter trade-offs to determine power's dependence on design choices.

As an alternative to simulation, recent work has proposed methods for estimating activity factors from hardware performance counters on live running systems [62, 114, 119]. Like simulation, such methods still draw on other estimations for capacitance and voltage. The key is that hardware performance counters can often serve as very accurate proxies for activity factor.

If one's goal is to measure aggregate power dissipation averaged over several cycles, then aggregate performance counters, such as instructions-issued-per-cycle, may offer surprisingly good estimations with few counters required. For example, Fan et al. used such IPC estimates to guide aggregate provisioning decisions in data centers [74]. Joseph and Martonosi used an early version of such techniques to estimate power on an Intel Pentium Pro microprocessor [119]. For subsequent microprocessors in which clock gating (and other techniques) mean more widely varying power, these approaches needed to be refined in terms of how individual performance counters were weighted and summed to provide an overall power estimate. Contreras and Martonosi describe one such approach with offline linear estimates created based on specially written benchmarks [62].

The techniques described above set up a relationship between total power and a sum of performance-counter-provided activity factors, each weighted to generate accurate total power values. While such approaches are good for tracking aggregate power as it varies in real time, they lack, however, the ability to discern where (i.e., in which units) the power dissipation is most or least prominent. Such per-unit power attribution is useful both for guiding power-efficient design optimizations, as well as for guiding thermal models of on-chip hotspots. For example, a total power estimate will merely tell me if I am near or exceeding the overall chip power budget or thermal capacity; it cannot tell me about whether I have one particular hotspot on the chip that is nearing its local thermal limit.

Accurate and efficient per-unit power estimators can be built by exploiting the specific hardware performance counters provided on nearly all high-performance microprocessors today. For example, Isci and Martonosi demonstrated an accurate counter-based estimator for the Intel Pentium 4 chips [114]. In this work, rather than aiming to provide a single total power estimate, they instead selected 22 physical hardware blocks from a die photo, and aimed to estimate the power of each of these units individually. Such floorplan-based per-unit estimates can be used to drive long-running thermal studies. For a particular hardware unit i, the power estimate is expressed as:

$$\text{Power}(i) = \text{AccessRate}(i) \times \text{ArchitecturalScaling}(i) \times \text{MaxPower}(i) \\ + \text{NonGatedPower}(i).$$

AccessRate can be measured or deduced via hardware performance counters. The other factors are determined by measurements using a set of benchmarks designed to isolate and exercise units of the hardware as independently as possible. Overall, for a wide variety of both SPEC benchmarks and desktop applications, their approach offers accuracy to within 2–4 W over the full operating range (roughly 5–55 W) of the Pentium 4 implementation they studied.

2.4.2   Imaging and Other Techniques

In recent years, interesting direct methods for measuring chip activity have emerged based on imaging technology. We briefly discuss these approaches here.

Thermal imaging: The central observation that drives this class of measurement techniques is that the thermal behavior of a running microprocessor can be observed, under the right conditions, as infrared (IR) radiation [91, 165]. The keys to such setups are in creating a system in which: (i) the observations can be made on a running chip and (ii) the infrastructure required to image the chip does not excessively perturb the running system.

The massive heatsinks currently used on microprocessors are clearly not infrared transparent. In addition, since they spread the heat the microprocessor gives off, they make it impossible to attribute hot spots in the IR image to particular localized hardware units. To prepare a chip for thermal measurements based on IR imaging, the chip must be operated without a conventional heat sink. Since this would normally cause the chip either to shut down (we hope!) or to malfunction (we fear!), an alternative method of cooling must be used. This alternative cooling is normally provided by pumping a liquid that is infrared transparent (such as mineral oil) over the microprocessor.

In addition to providing dynamic cooling for the chip, the measurements also can benefit if the chip can be thinned in order to provide a more direct imaging path to the active silicon layer. That is, due to the bonding techniques typically used today, the active silicon layer lies face down in the socket, with the thicker silicon wafer material above it. Thinning down this material allows the chip to still function while allowing the imaging equipment to capture the heating effects without spreading.

Power estimates from thermal images: Previously, we described a methodology in which hardware performance counters first drove a power measurement from which some thermal estimates could be deduced. Here, an opposite approach applies. Namely, from the thermal imaging techniques just described, one can garner good still photos or videos of running chips in which the color of the IR image corresponds to temperature. One can provide an intuitive or qualitative sense of hotspots simply by overlaying these IR images with a floorplan of the chip: units that are chip hotspots will be colored red [91, 165].

From these IR images, one can also provide more quantitative data regarding chip temperatures and power dissipation. For example, if there are small on-chip digital thermometers, then one can calibrate between thermal readings at a particular point on the chip versus the detailed image data for that point on the chip. Such calibrations allow one to provide image processing filters that calibrate the camera and the setup to provide accurate absolute (not simply relative) temperature data.

In addition, one can back-calculate from this temperature to deduce the per-unit power dissipation that must have led to them. For example, Mesa-Martinez et al. used an "inverse heat transfer" solution based on genetic algorithms to "match" the collected thermal images back to a consistent model for how per-unit power dissipation must have varied in order to produce those images [165].

2.5   SUMMARY

Overall, this chapter has provided an introduction to the aspects of CMOS power consumption that are relevant to computer architecture in current and next-generation designs. By summarizing metrics, models, and simulation techniques, we have offered the groundwork from which subsequent chapters can discuss power optimization techniques themselves.

CHAPTER 3

Using Voltage and Frequency Adjustments to Manage Dynamic Power

Issues addressing dynamic power have predominated the power-aware architecture landscape. Amongst these dynamic power techniques, methods for addressing voltage and frequency have dominated in turn. Most of these methods have focused on dynamic adjustments to supply voltage, clock frequency, of both, and they go under the broad title of *Dynamic Voltage and Frequency Scaling*, or *DVFS*. This chapter discusses the motivation for these techniques overall, and gives examples drawn from different categories of techniques.

Chapter Structure: Decisions to engage voltage and frequency scaling are made at various levels. The decision level, the level of the control policy, defines the structure of this chapter. Starting from the top, the system (or operating system) level, the chapter unfolds to progressively more focused levels: program (or program phase) level and the hardware (flip-flop) level. The following section (Section 3.1) gives an overview of voltage/frequency scaling and discusses a number of issues pertaining to the corresponding techniques.

3.1 DYNAMIC VOLTAGE AND FREQUENCY SCALING: MOTIVATION AND OVERVIEW

The basic dynamic power equation: $P = CV^2 Af$ clearly shows the significant leverage possible by adjusting voltage and frequency [47, 101]. If we can reduce voltage by some small factor, we can reduce power by the square of that factor. Reducing supply voltage, however, might possibly reduce the performance of systems as well. In particular, reducing supply voltage often slows transistors such that reducing the clock frequency is also required. The benefit of this is that within a given system, scaling supply voltage down now offers the potential of a cubic reduction in power dissipation. The downside of this is that it may also linearly degrade performance. If the program runs at lower power dissipation levels, but for longer durations, then the benefit in terms of total energy will not be cubic. It is interesting to note that while voltage/frequency scaling improves EDP (because the reduction in power outpaces the reduction in performance), it can do no better than break even on the ED^2P metric (and this, only when the scaling factors for frequency and voltage are the same).

Nonetheless, DVFS is appealing first because max-power limits may welcome max-power reductions even if the energy is not reduced much. In addition, DVFS is appealing because often we can discern ways, as this chapter will discuss, to reduce clock frequency without having the workload experience a proportional reduction in performance.

3.1.1 Design Issues and Overview

From an architect's perspective, key design issues for DVFS include the following:

(1) At what level should the DVFS control policies operate? Fundamentally, DVFS approaches exploit *slack*. Slack can appear at different levels and various DVFS approaches have been proposed for each level. Approaches operating at the same level share a similar set of mechanisms, constraints, and available information. We can discern three major levels where DVFS decisions can be made:

- *System-level based on system slack*: At this level, the *idleness* of the whole system is the factor that drives DVFS decisions (Section 3.2). In many cases, decisions are taken according to system load. The *whole* processor (or embedded system, wireless system, etc.) is typically voltage/frequency scaled to eliminate idle periods.

- *Program- or program-phase-level based on instruction slack*: Here, decisions are taken according to program (or program phase) behavior (Section 3.3 for a single clock domain and Section 3.4 for multiple clock domains). *Instruction Slack* due to long-latency memory operations is typically exploited at this level for DVFS in single-threaded programs. In multi-core processors, the ability to run parallel (multi-threaded) programs opens up the possibility for the *parallel* program behavior to drive voltage/frequency decisions.

- *Hardware-level based on hardware slack*: Finally, a recent approach, called *Razor*, goes below the program level, right to the hardware (Section 3.5). Razor tries to exploit *slack* hidden in hardware operation. This slack exists because of *margins* needed to isolate each hardware abstraction layer from variations in lower levels. This slack is exploited similarly to the way idle time is exploited at the system level.

(2) How will the DVFS settings be selected and orchestrated? In some cases, DVFS approaches may allow software to adjust a register which encodes the desired (V, f) setting. In other cases, the choices will be made dynamically "under the covers" by hardware mechanisms

alone. In either scenario, research questions arise regarding whether to make offline (e.g., compile-time) decisions about DVFS settings, versus online, reactive, approaches.

(3) What is the hardware granularity at which voltage and frequency can be controlled? This question is closely related to the question above. The bulk of the DVFS research has focused on cases in which the entire processor core operates at the same (V, f) setting but is asynchronous to the "outside" work, such as main memory. In such scenarios, the main goal of DVFS is to capitalize on cases in which the processor's workload is heavily memory-bound. In these cases, the processor is often stalled waiting on memory, so reducing its supply voltage and clock frequency will reduce power and energy without having significant impact on performance.

Other work has considered cases in which multiple clock domains may exist on a chip. These so-called MCD scenarios might either be multiple clock domains within a single processor core [199, 200, 216, 227, 228] or chip multiprocessors in which each on-chip processor core has a different voltage/clock domain [67]. This dimension is explored in Section 3.4.

(4) How do the implementation characteristics of the DVFS approach being used affect the strategies to employ? Some of the implementation characteristics for DVFS can have significant influence on the strategies an architect might choose, and the likely payoffs they might offer. For example, what is the delay required to engage a new setting of (V, f)? (And, can the processor continue to execute during the transition from one (V, f) pair to another?) If the delay is very short, then simple reactive techniques may offer high payoff. If the delay is quite long, however, then techniques based on more intelligent or offline analysis might make more sense.

(5) How does the DVFS landscape change when considering parallel applications on multiple-core processors? When considering one, single-threaded application in isolation, one need only consider the possible asynchrony between compute and memory. In other regards, reducing the clock frequency proportionately degrades the performance. In a parallel scenario, however, reducing the clock frequency of one thread may impact other dependent threads that are waiting for a result to be produced. Thus, when considering DVFS for parallel applications, some notion of critical path analysis may be helpful.

Another similar question regards whether continuous settings of (V, f) pairs are possible, or whether these values can only be changed in fixed, discrete steps. If only discrete step-wise adjustments of (V, f) are possible, then the optimization space becomes difficult to navigate because it is "non-convex." As a result, simple online techniques might have difficulty finding global optima, and more complicated or offline analysis again becomes warranted.

Because DVFS is available for experimentation on real systems [111, 112, 2], and because it offers such high leverage in power/energy savings, it has been widely studied in a variety of communities. Our discussion only touches on some of the key observations from the *architectural* research arena, but we urge readers to explore research ideas from the design automation and other communities as well.

3.2   SYSTEM-LEVEL DVFS

3.2.1   Eliminating Idle Time

Architectural techniques for dynamic voltage and frequency scaling first appeared in the literature pertaining to the system (or operating system) level. Commercial implementations controlled at this level are also the most common form of DVFS (e.g., Intel's Enhanced SpeedStep and AMD's PowerNow!^{TM}).

Wiser, Welchm, Demers, and Shenker, all of Xerox PARC, first published on this type of DVFS [223]. Wiser et al. observed that *idle time represents energy waste*. To understand why this is, consider the case of a processor finishing up all its tasks well within the time of a scheduling quantum. The remaining time until the end of the quantum is *idle time*. Typically, an idle loop is running in this time but let us assume that the processor can be stopped and enter a sleep mode during this time. One could surmise that a profitable policy would be to go as fast as possible, finish up all the work and then enter the sleep mode for the idle time and expend little or no energy. But that is not so.

As an example, let us assume that the time needed to finish up all the work is half the time quantum. The idle time would then be the other half. At best, if the sleep mode wastes no energy, *half* the energy that would be expended in a busy quantum can be saved in this way. Consider now the case where we have the ability to dynamically scale both frequency *and* voltage. Just by slowing down the clock, the work that needs to be done can be stretched to cover the entire quantum. In our example, assuming that the clock period is doubled (frequency is halved) to eliminate the idle time, power consumption drops by half. Thus, the *energy* needed to complete the work is the same as going full speed for half the quantum and idling for the rest. But, on top of that we can also scale the voltage which reduces power consumption quadratically. The end result, with DVFS, is that the more we stretch a fixed amount of work the more we gain, thus turning idle time into *opportunity cost*.[1]

With this motivation, Wiser et al. propose three scheduling algorithms, called OPT, FUTURE, and PAST, aiming to eliminate idle time. Their work specifically targets idle time as it is experienced in the operating system, i.e., the time taken by the idle loop, or I/O waiting time. Of course, when one considers very long idle periods (e.g., periods measured in seconds) and includes components such as the display or the disk of a portable system in the accounting for total power, the best policy is to shut down all components (since the display and disk surpass

---

[1]An important point here is that *static* power consumption is not taken into account in this reasoning. We will return to this in Chapter 5 where we discuss the implications of static power in relation to DVFS.

the processor in power consumption). In this situation, the scheduling algorithms described here may not be applicable.

The scheduling algorithms studied by Wiser et al. are meant to minimize the time spent in the system's idle loop for short bursts of idle activity. Instead of actually implementing these algorithms in a real system, Wiser et al. collected traces and used them to model the effects on the total power consumption of the processor. The traces contain timestamps of context switches, entering and exiting the system idle loop, process creation and destruction, and waiting or waking up on events. They come from workstations running a variety of different workloads, such as software development and other typical engineering tasks. To prevent whole system shut-down (processor, display, disk), any period of 30s or longer with a load below 10% was excluded from consideration.

All three scheduling algorithms are interval-based. Traces are divided into fixed-length intervals, and the proportion of time that the CPU is active within each interval is computed individually. At the end of each interval, the speed of the processor for the upcoming interval is decided. The goal is to minimize—eliminate if possible—idle time. If, however, the quantum deadline is missed, i.e., the processor cannot finish its assigned work within the quantum limits, any unfinished work spills over to the next quantum. From the three scheduling algorithms, the first two are impractical since they can look into the future of the trace data, while the third is a plausible candidate for implementation.

OPT, FUTURE, and PAST: OPT is a simplified Oracle algorithm that perfectly eliminates idle time in every quantum by stretching the run times in a trace. It can look arbitrarily far into the future. It provides a reference point for scheduling all work in a power-optimal way. However, it makes several over-simplifications. First, it does not make a distinction between "soft" and "hard" idle time. The hard idle time is necessary waiting (e.g., for I/O) that should not be stretched or compressed. In addition, it does not care on how long a job is delayed, as long as it finishes by the end of the trace. This may result in very slow response times especially for the interactive jobs.

FUTURE is a simple modification of OPT that can only look into the subsequent interval. The repercussion of this choice is that no work is delayed past the end of the next interval. For large intervals, FUTURE approaches OPT in terms of energy savings, while for smaller ones it falls behind. Like OPT, FUTURE is also unrealistic for an on-line implementation, since it still peeks into the future.

The PAST algorithm, which is the only one of the three suitable for an on-line implementation, looks into the past in order to predict the future. As with the previous algorithms, its interval size can be adjusted for different results. PAST works under the assumption that the next interval is similar to the current one. Although this may seem naive, PAST stands up quite well even compared to newer, more sophisticated, scheduling algorithms [89].

Based on the current interval, PAST assesses the number of cycles that the processor is going to be busy in the next interval. If the processor, because of its speed setting, misses the deadline to complete its work in the current interval, unfinished work spills over to the next interval. If, on the other hand, the processor completes its work before the end of the quantum, the remaining idle time is taken into account for the speed setting for the next interval. The speed setting policy raises speed if the current interval was more busy than idle and lowers speed if idle time exceeds some percentage of the quantum time. These comparisons (busy versus idle, as a fraction of the quantum) are based on empirically derived parameters which lead to speed changes that smooth the transitions from high to low frequencies.

Wiser et al. examine several voltage minima and several interval sizes in relation to the three algorithms. PAST tends to fall behind when a light-load interval is followed by a heavy-load interval. Unfinished work spills over to the next interval causing speed to vary more from interval to interval until PAST manages to catch up. Because of this, it is less efficient in power consumption than either OPT or FUTURE.

In general, there is a trade-off between the number of missed deadlines and energy savings which depends on interval size. The smaller the interval, the fewer the missed deadlines because speed can be adjusted at a finer time resolution. But energy savings are smaller because there is a frequent switching between high and low speeds. In contrast, with large intervals, better energy savings can be achieved, but at the expense of more missed deadlines, more work spilled-over, and, as a result, a decreased response time for the workload. Regarding actual results, Wiser et al. conclude that, for their setup, the optimal interval size ranges between 20 and 30 ms yielding power savings between 5% and 75%.

3.2.2   Discovering and Exploiting Deadlines

Whereas the DVFS techniques of Wiser et al. are based on the idle time as seen by the operating system (OS) (e.g., the idle loop), Flautner, Reinhardt, and Mudge look into a more general problem on how to reduce frequency and voltage without missing deadlines [78]. Their technique targets general purpose systems that run interactive workloads.

What do "deadlines" mean in this context? In the area of *real-time* systems, the notion of a deadline is well defined. Hard real-time systems have fixed, *known* deadlines that have to be respected at all times. Since most real time systems are embedded systems with a well-understood workload, they can be designed (scheduled) to operate at an optimal frequency and voltage, consuming minimum energy while meeting all deadlines. An example would be a mobile handset running voice codecs. If the real-time workload is *not* mixed with non-real-time applications, then DVFS controlled by an on-line policy is probably not necessary—scheduling can be determined off-line.

Flautner et al. consider an entirely different class of machines. In general-purpose machines running an operating system such as Linux, program deadlines have to do more with user perception than with some strict formulation. Thus, the goal in their work is to discover "deadlines" in irregular and multiprogrammed workloads that ensure the quality of interactive performance.

The approach to derive deadlines is by examining communication patterns from within the OS kernel. Application interaction with the OS kernel reveals the, so-called, *execution episodes* corresponding to different communication patterns. This allows the classification of tasks into interactive, periodic producer, and periodic consumer. Depending on the classification of each task, deadlines are established for their execution episodes. In particular, the execution episodes of interactive tasks are assigned deadlines corresponding to the user-perception threshold, which is in the range of 50–100 ms. Periodic producer and consumer tasks are assigned deadlines corresponding to their periodicity. All this happens within the kernel without requiring modification of the applications.

Having a set of deadlines for the interactive and the periodic tasks, frequency and voltage settings are then derived so that the execution episodes finish within their assigned deadlines. The approach can result in energy savings of 75% without altering the user experience [78].

3.3   PROGRAM-LEVEL DVFS

Dropping deeper, the program and program-phase level includes all DVFS techniques that apply to whole programs or program phases. These include both offline techniques with analysis performed by compilers, as well as online and phase-based techniques.

3.3.1   Offline Compiler Analysis

DVFS control is exposed to the software level through instructions that can set particular values of (V, f). These mode-set instructions are provided in nearly all contemporary microprocessors, including Intel XScale [112] and AMD Mobile K6 Plus [2]. However, the use of these instructions has been largely at the process/task level under operating system control. More recent work has studied program-aware DVFS in which adjustment code is inserted into the application executable itself.

Because DVFS adjustments incur some time and energy cost each time they are applied, it is tempting to use offline analysis techniques to carefully plan and optimize their invocation. The goal is to identify regions for which (V, f) adjustments can be helpful, and, where possible, to coarsen these granularities in order to amortize the overheads of DVFS adjustment.

Much work has been done on scheduling DVFS adjustments. We focus here on several of the more "applied" works.

Profile-assisted compiler approach: Hsu and Kremer's work provided a heuristic technique that lowers the voltage for memory-bound sections [103]. The intuition behind their approach is that if the processor and memory operate largely asynchronously from each other, then the processor can be dialed down to much lower clock frequencies during memory-bound regions, with considerable energy savings but no significant performance loss. They implemented their technique within the SUIF2 source-to-source compiler infrastructure (gcc compilers were used to generate object code).

The compiler algorithm is based on heuristics and profiling information to solve a minimization problem. Using the author's description, the problem can be stated as follows:

Given a program P, find a program region R and a frequency f (lower than the maximum frequency f_{max}) such that, if R is executed at the reduced frequency f and with reduced voltage,

- the total execution time (including the voltage/frequency scaling overhead) is not increased more than a small factor over the original execution time, and

- the total energy usage is minimized.

Candidate regions are considered to be loop nests, call sites, called procedures, statement sequences (straight-line code), or even the entire program. Restricting regions to the above programming constructs has the benefit of making the number of DVFS switchings tractable, since the number of times such regions execute can be determined with reasonable accuracy either statically or by *profiling*. DVFS occurs only on entering and exiting a region. Finally, candidate regions are selected by size, so DVFS switchings occur only for significantly large pieces of code.

To implement a compiler algorithm to solve this minimization problem, two pieces of information are needed for each candidate region R: an estimate of its execution time at a frequency f, denoted $T(R, f)$, and the number of times $N(R)$ the region executes during the lifetime of the program. $T(R, f)$ and $N(R)$ are computed, depending on the programming construct involved, according to the rules shown in Figure 3.1.

$T(R, f)$ values for regions, that do not decompose further into smaller regions, are provided by profiling—along with the $N(R)$ values that cannot be computed statically.[2] Using the $T(R, f)$ and $N(R)$ information, compiler heuristics then select the appropriate regions to annotate for DVFS.

Hsu and Kremer use an experimental setup to measure power in laptops (with Linux and GNU compilers). With the help of a digital power meter and by annotating the programs with mode-set instructions, which select DVFS settings on AMD mobile Athlon 4 and Transmeta

---

[2]The authors cite analytic techniques to compute $T(R, f)$ given information for $T(R, f_{max})$, but these techniques were not used in practice.

```
if statement:
R: if() then R₁ else R₂
       T(R,f) = T(R₁,f)+T(R₂,f)
       N(R) = N(R₁)+N(R₂)

explicit loop structure:
R: loop()R₁
       T(R,f) = T(R₁,f)
       N(R) is profiled

call site:
R: call F()
       T(R,f) = T(F,f)*N(R)/N(F)
       N(R) is profiled

sequence of regions:
R: sequence(R₁,...,Rₙ)
       T(R,f) = ?{T(Rᵢ,f): 1<=i<=n}
       N(R) = N(R₁) = ... = N(Rₙ)

procedure:
F: procedure F() R
       T(R,f) = T(?,f)
       N(R) = ?{?(Rᵢ): Rᵢ is a call site to F()}
```

FIGURE 3.1: Rules for deriving execution time and number of invocations for regions. Adapted from [103].

Crusoe processors, they show concrete results for their compiler technique for *total system power*. Hsu and Kremer report energy savings of up to 28% with performance degradation of less than 5% for the SPECfp95 benchmarks.

Analytic techniques: While heuristic techniques offer some benefits, subsequent work has sought to refine these techniques toward optimal or bounded-near-optimal solutions. For example, subsequent work by Saputra et al. provided an exact mixed-integer linear programming (MILP) technique that can determine the appropriate (V, f) setting for each loop nest [197]. An MILP approach is required because discrete (V, f) settings lead to a non-convex optimization space. Their technique reports improvements in energy savings compared to prior work. However, it does not account for the energy penalties incurred by mode switching. Furthermore, the long runtimes of straightforward MILP approaches make their integration into a compiler somewhat undesirable.

Work by Xie et al. expanded on these ideas in several ways [229, 230]. First, they expanded the MILP approach by including energy penalties for mode switches, providing a much finer grain of program control, and enabling the use of multiple input data categories to determine optimal settings. In addition, they determined efficient methods for solving the MILP optimization problem with boundable distance from the true optimal solution. While the time and energy savings offered by the MILP approach vary heavily depending on the application performance goal and the (V, f) settings available, as much as 2X improvements have been shown in some cases.

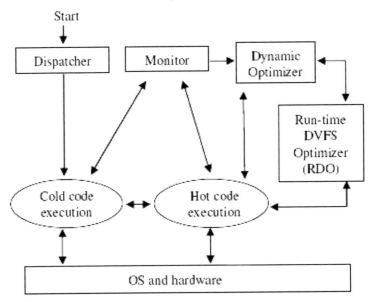

FIGURE 3.2: Dynamic compilation system. Reproduced from [226]. Copyright 2005 IEEE.

3.3.2   Online Dynamic Compiler analysis

While offline compiler analysis can use global program knowledge and detailed analysis to plan DVFS adjustments, it lacks knowledge about runtime characteristics such as data inputs and caching behavior that can greatly impact program behavior. Thus, much work has been done on more dynamic techniques for determining where to place DVFS adjustments.

In a more recent work by Wu et al. the authors studied methods using dynamic compilation techniques to analyze program behavior and also to dynamically insert DVFS adjustments at the locations determined to be most fruitful [226]. Wu et al. implemented a prototype of this runtime DVFS optimizer (RDO) and integrated it into an industrial-strength dynamic optimization system (a variant of the Intel PIN system [159]). A block diagram of their approach is shown in Figure 3.2.

The dynamic optimizer begins by dispatching "cold" code for execution. A monitor determines whether this code is frequently executed. In this case, the RDO optimization is applied along with other conventional performance optimizations of the dynamic optimizer.

The RDO flowchart is depicted in Figure 3.3. The first order of business is to determine the "hotness" of functions and first-level loops in main(). These code regions are instrumented for frequency-of-execution profiling. If they become "hot," they are then considered for DVFS optimization. To decide on the optimization, an analytic model determines whether a region is memory-bound or CPU-bound. If the region is clearly memory-bound, it is instrumented with DVFS mode-set instructions. This is because memory-bound code has enough *slack* so that

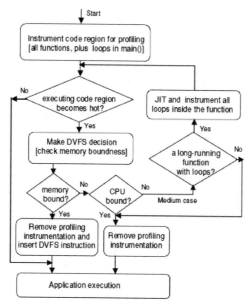

FIGURE 3.3: RDO flowchart. Reproduced from [226]. Copyright 2005 IEEE.

slowing it down does not hurt the overall performance and can save power. DVFS approaches, after all, are exploiting slack. If the code is CPU-bound, it is left alone since slowing it down could seriously degrade the performance. If a decision for memory- or CPU-boundedness cannot be made and the region is large enough, it is divided up into smaller regions and the algorithm repeats for each of the smaller regions.

The analytic model determines slack in the CPU due to memory operations. The model divides CPU execution into execution that happens *concurrently* with outstanding memory operations and execution that *depends* on memory operations (Figure 3.4). The concurrent execution, hidden behind the latency (t_{asym_mem}) of a memory operation can be *stretched* to take up slack. The more a code region is characterized by this type of execution the slower it can run without affecting the end performance too much. At run-time, the analytic model is approximated using hardware performance counters. These counters provide information on the number of retired micro-operations (μops) per memory bus transaction (which relates to the memory latency t_{asym_mem}), and the number of completed instructions during outstanding memory (operations which relates to the portion of concurrent execution) [226].

Power measurements were taken on an actual system using RDO on a variety of benchmarks. The experimental setup consisted of a voltage/current measurement unit (measuring the voltage and current drawn by the target processor), a signal conditioning unit (to reduce noise in measurements), a data acquisition unit, and a second computer acting as a data logging and

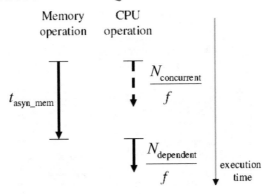

FIGURE 3.4: Slack in CPU due to memory operations. Rproduced from [226]. Copyright 2005 IEEE.

processing unit. On average, their results achieved an energy-delay product (EDP) improvement (over non-DVFS approaches) of 22.4% for SPEC95 FP, 21.5% for SPEC2K FP, 6.0% for SPEC2K INT, and 22.7% for Olden benchmarks. These represent three to five times better results than a baseline approach based on static DVFS decisions.

3.3.3   Coarse-Grained Analysis Based on Power Phases

The previously discussed compiler approaches used detailed off-line or on-line program analysis to discern useful DVFS adjustment points. The online techniques of Wu et al. achieved their detailed program knowledge through relatively high-overhead dynamic monitoring. Thus, it is tempting to look for techniques that maintain such detailed knowledge but reduce monitoring overhead. Since most general-purpose processors include a suite of user-readable hardware performance counters, it is possible to build up a history of program behavior from seeing aggregate event counts.

In particular, early work by Isci and Martonosi demonstrated how these event counts can be viewed as identifying "fingerprints" of program phase behavior [115]. Essentially, this work aggregated power data based on different hardware counters into a summation of different power subcomponents. If each subcomponent is treated as one dimension in a vector space, then these so-called power vectors can be used to identify unique aspects of power behavior that call for different management approaches.

More recently, Isci, Contreras, and Martonosi elaborated on their technique by including a predictor table that can predict future power behavior based on recently observed values [116]. This so-called Global Phase History Table (GPHT) is inspired by hardware branch predictors, but is implemented in software by the operating system. Like a branch predictor, it stores a "history table" of recently measured application metrics that are predictive of proper DVFS adjustments. For example, one prototype implementation measured "memory operations per micro-operation," and used this as an indicator of the memory boundedness indicative of likely DVFS effectiveness. For each pattern of past behavior stored in a history entry, a different prediction of next-step behavior can be made. For each next-step prediction, there is a one-to-one mapping to an appropriate DVFS setting. If the DVFS setting is different from the current setting, then the V, f are adjusted accordingly. When guided by the GPHT, DVFS was found to achieve EDP improvements as high as 34% for the highly variable benchmarks that this approach targets.

3.4   PROGRAM-LEVEL DVFS FOR MULTIPLE-CLOCK DOMAINS

Some of the early architectural work on DVFS actually focused on opportunities within multiple-clock-domain (MCD) processors. The rationale for MCD processors is that as feature sizes get smaller, it becomes more difficult and expensive to distribute a global clock signal with low skew through the processor die. Thus, researchers have explored globally-asynchronous locally-synchronous (GALS) techniques.

Scaling voltage/frequency independently for each clock domain within a processor can be done dynamically (Section 3.4.1) or statically (Section 3.4.2); both cases aim to exploit *slack* in the execution of individual instructions.

Finally, the emerging architectural paradigm for deep sub-micron technologies, the multi-core paradigm, can be considered as an MCD design where synchronous cores operate asynchronously to each other. DVFS techniques for multi-cores are discussed in Section 3.4.3.

3.4.1   DVFS for MCD Processors

In GALS approaches, a processor core is divided into synchronous islands, each of which is then interconnected asynchronously but with added circuitry to avoid metastability. The islands are typically intended to correspond to different functional units, such as the instruction fetch unit, the ALUs, the load-store unit, and so forth. A typical division is shown in Figure 3.5.

In early GALS DVFS work, Marculescu and her students considered the performance and power implications of GALS designs [216, 117]. In [117], they first predicted that going from a synchronous to a GALS design caused a drop in performance, but that elimination of the global clock would not single-handedly lead to drastic power reductions. In fact, from a power perspective, GALS designs are initially less efficient when compared to synchronous architectures. Their potential, however, lies in the flexibility offered by having several independently controllable clocks. As with other DVFS opportunities, the key lies in finding inter-domain slack that one can exploit. For example, in some MCD designs, the floating point unit could be clocked much more slowly than the instruction fetch unit, because its throughput and latency demands are lower. Iyer and Marculescu's results show that for a GALS processor with five

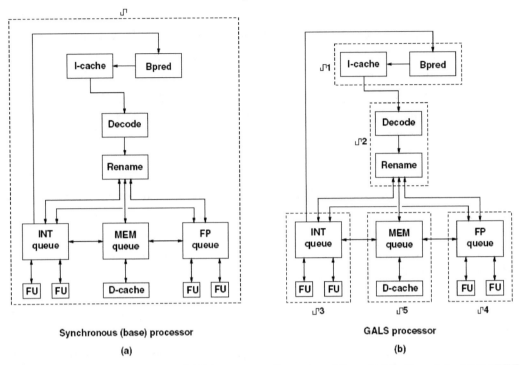

FIGURE 3.5: Synchronous versus GALS processor. Reproduced from [117]. Copyright 2002 IEEE.

clock domains, the drop in performance ranges between 5% and 15%, while power consumption is reduced by 10% on the average. Thus, fine-grained voltage scaling allows GALS to match or exceed the power efficiency of fully synchronous approaches.

In a similar timeframe, research from Albonesi's group also explored DVFS opportunities in MCD processors [199, 200]. Similar to the Iyer and Marculescu MCD division of the CPU (Figure 3.5), Semeraro et al. divide the processor into five domains: Front end, Integer, Floating point, Load/Store, and External (Main Memory). The division is shown in Figure 3.6 along with the relevant clock parameters. The domains interface via queues.

The work by Semerao et al. primarily focused on the control policies by which (V, f) settings could be optimized for such microarchitectures. Their early work used offline scheduling to accomplish good (V, f) settings, but subsequent work explored control-theoretic approaches for managing MCD processors.

Offline approach: Semeraro et al. use an offline approach in [200] to select the times and frequency values for DVFS in an application. The application is executed (at maximum speed) in a simulator that creates an event trace. The events correspond to primitive operations in the processor (for example, for a load instruction: fetch, dispatch, address calculation, memory access, and commit events are traced). The events are connected with resource constraints and

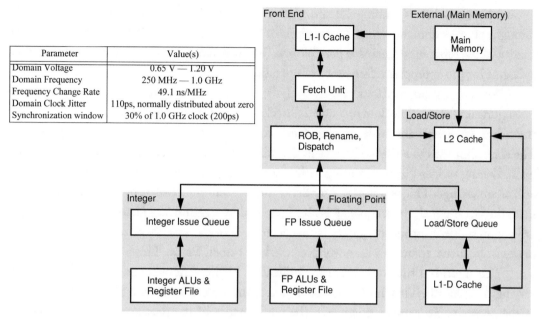

Parameter	Value(s)
Domain Voltage	0.65 V — 1.20 V
Domain Frequency	250 MHz — 1.0 GHz
Frequency Change Rate	49.1 ns/MHz
Domain Clock Jitter	110ps, normally distributed about zero
Synchronization window	30% of 1.0 GHz clock (200ps)

FIGURE 3.6: MCD processor and clock parameters. Adapted from [199].

data dependencies into a temporal-ordered directed acyclic graph (DAG). A DAG is created for an interval of 50K instructions and is then processed in two phases.

In the first phase, each event in the DAG that is not on the critical path is stretched, as if each instruction could run at its own frequency. A multi-pass "shaker" algorithm tries to distribute slack evenly in the DAG wherever it exists. This step concludes when all slack in the DAG is removed and each instruction is assigned to run at one of the allowed frequencies (e.g., one of the 32 frequencies for the Transmeta Crusoe or one of the 320 in the Intel XScale).

Since executing each instruction at a different frequency is not practical, the second phase processes the results of the first phase and aims to find a single *minimum* frequency *per interval for each domain*. This is done under the constraint that each domain finishes its work with no more than a fixed—externally set—factor of time dilation. Finally, intervals with the same or similar frequencies are merged together to create larger combined intervals—and this is continued recursively—with the intent of reducing the number of reconfigurations. In contrast to the first phase where DVFS reconfiguration was considered instantaneous and for-free, in the second phase realistic DVFS overhead (for the two processors studied) is taken into account.

Online approach: By analyzing the resource utilization of various CPU structures in [199], Semeraro et al. discovered that there is a significant correlation between the number of valid entries in the input queues for each domain and the desired frequency of the domain as derived by their offline method (see above). In other words, the *occupancy* of these queues reveals the throughput of the corresponding domain.[3] When a queue fills up—its utilization increases—the throughput of its domain is low. The analysis of Semeraro et al. revealed that (i) decentralized control of the different domains is possible and (ii) the utilization of the input queues is a good indicator for the appropriate frequency of operation.

Based on the observations of their analysis, Semeraro et al. devised an online DVFS control algorithm for multiple domains called Attack/Decay. This is a decentralized, interval-based algorithm. Decisions are made independently for each domain at regular sampling intervals. The algorithm tries to react to changes in the utilization of the issue (input) queue of each domain. During sudden changes, the algorithm sets the frequency aggressively to try to match the utilization change. This is the *Attack* mode. If the utilization increased by a significant amount since the last interval, the frequency is also increased by a significant factor. Conversely, when utilization suddenly drops, frequency is also decreased. In the absence of any significant change in the issue queue, frequency is slowly decreased by a small factor. This is the *Decay* mode.

The algorithm tries to balance the utilization of the various domains independently by varying their speeds. This, however, does not account for a natural change in the performance of the program. To capture this effect, the performance of the processor in terms of IPC is tracked from interval to interval. If there is an inherent change in the IPC that is not related to frequency adjustments in the domains, then no frequency changes are allowed for the next interval. The IPC is the only global information needed in this algorithm.

On average, across a wide range of MediaBench, Olden, and Spec2000 benchmarks, their algorithm achieved a 19% reduction (from a non-DVFS baseline) in energy-per-instruction and a 16.7% improvement in energy-delay product. The approach incurred a modest 3.2% increase in cycles per instruction (CPI). Interestingly, their online control-theoretic approach was able to achieve a full 85.5% of the EDP improvement offered by the prior offline scheduling approach. Wu et al. extended the online approach using formal control theory and a dynamic stochastic model based on input-queue occupancy for the MCDs [228].

3.4.2   Dynamic Work-Steering for MCD Processors

As an alternative to dynamic voltage/frequency scaling of multiple clock domains, one can *statically* provide multiple components for the same function clocked at different frequencies. For example, one can provide a fast and power-hungry pipeline and a slow but power-efficient pipeline. With two such pipelines, the problem is no longer about selecting domain frequencies to eliminate stack, but instead is about steering instructions to the appropriate slow or fast

---

[3]The occupancy of the instruction queue is also used for resizing it to reduce its switching activity factor as we discuss in Section 4.6.3.

pipeline to accomplish this. This is a *work-steering* strategy that is also revisited in Chapter 4 for *effective capacitance* optimizations (Section 4.13).[4]

Fields, Bodik, and Hill use this work-steering approach as an example of how instruction slack can be exploited [76]. In their study, they show that there exists a significant slack in instructions. In many instances, instructions can be delayed several cycles without any impact on the program's critical path and hence its performance [76]. Furthermore, they classify the instruction slack into *local*, *global*, and *apportioned*. *Local* slack exists when an instruction can be delayed without any impact on any other instruction. *Global* slack exists when delaying an instruction does not delay the last instruction of the program (i.e., there is no impact on the total execution time). *Apportioned* slack refers to the amount of slack for a *group of instructions* that can be delayed together without impact on execution. Apportioned slack depends on how it is calculated from the instructions' individual slack [76].

To measure slack, Fields et al. use an offline analysis (similar to the Semeraro et al. offline approach used in [200] and described in Section 3.4.1) that creates a dependence graph of the execution taking into account both data dependencies and microarchitectural resource constraints. Their offline approach allows the calculation of all three types of slack. Their results show that there is enormous potential for exploiting slack by slowing down instructions [76].

More interestingly, Fields et al. show that one can dynamically predict slack in hardware. This is of significance since it allows for the possibility of *fine-grain*—on a per-instruction basis—control policies. Online control policies discussed previously for DVFS in MCD processors cannot treat each instruction individually. There is simply no possibility of dynamically changing the frequency of execution individually for each instruction; instead, the frequency of each domain is adjusted according to the aggregate behavior of all the instructions processed in this domain over the course of a sampling interval (Section 3.4.1).

With work steering, the execution frequencies are fixed for each execution pipeline—as is the case for the fast and slow pipelines in Figure 3.7—and the instructions are steered toward the appropriate pipeline. All that is needed to implement work steering is to have a good idea of the slack of each instruction. And this is where prediction comes into play. According to Fields et al., for 68% of the static instructions, 90% of their dynamic instances have enough slack to double their latency. This slack "locality" allows slack prediction to be based on sparsely sampling dynamic instructions and determining their slack.

Slack prediction would not be feasible if slack could not be measured efficiently at run-time. To determine whether an instruction has slack, Fields et al. employ an elegant delay-

---

[4]There too, multiple components are provided, offering a range of power/performance characteristics, and work (computation) is dynamically steered according to run-time conditions and goals.

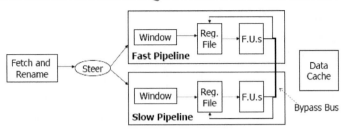

FIGURE 3.7: Work steering for a fast and a slow pipeline. Reproduced from [76]. Copyright 2002 IEEE.

and-observe approach. An instruction is delayed for a number of cycles and then observed to see if it becomes critical.

Criticality, in turn, is tested with a token that is passed from instruction to instruction, starting from the delayed instruction, and passed to all dependent instructions. The token is *dropped* if it is not passed to an instruction at the very last moment before it becomes ready to execute. In other words, the token is dropped if it has slack. If the token is still in existence well after an instruction is delayed, then the instruction, thus far, is in the critical path.

Slack determined by sampling is stored, per-instruction, in a PC-indexed predictor. This prediction is used in subsequent dynamic instances of the instruction for steering. If an instruction's predicted slack can accommodate the increased latency of a slow pipeline, then the instruction can be steered to this pipeline for execution without an impact on the performance. Fields et al. confirm this in their results, showing that a control policy based on slack prediction is second-best, in terms of performance, only to the ideal case of having two fast pipelines instead of a fast and a slow pipeline [76]. However, execution in the slow pipeline yields significant benefits in power consumption.

3.4.3   DVFS for Multi-Core Processors

As chip multiprocessors (CMPs) become the predominant general-purpose, high-performance microprocessor platform, it becomes important to consider how DVFS management can be applied to them most effectively. One major design decision concerns whether to apply DVFS at the chip level or at the per-core level. As with other MCD designs, per-core DVFS is considered more expensive; it requires more than one power/clock domain per chip, and synchronizer circuits are required to avoid metastability between domains. On the other hand, multiple clock domains may be employed anyway for circuit design or reliability reasons, in addition to voltage and frequency control.

Research has explored the benefits of per-core versus per-chip DVFS for CMPs. For example, on a four-core CMP in which DVFS was being employed to avoid thermal emergencies (rather than simply to save power), a per-core approach had 2.5× better throughput than a per-chip approach [67]. This is because the per-chip approach must scale down the entire chip's (V, f) when even a single core is nearing overheating. With per-core control, only the core with a hot spot must scale (V, f) downwards; other cores can maintain high speed unless they themselves encounter thermal problems.

While a multi-core processor can be used to run independent programs for throughput, its promise for single-program performance lies in thread-level parallelism. Managing power in a multicore when running parallel (multi-threaded) programs is currently a highly active area of research. Many research groups are tackling the problem, considering both symmetric architectures which replicate the same core and asymmetric architectures that feature a variety of cores with different power/performance characteristics [146].[5] Independent DVFS for each core [15], a mixture chip-wide DVFS and core allocation [153], or work-steering strategies at the program level in heterogeneous architectures [146, 170] are considered.

3.5   HARDWARE-LEVEL DVFS

The main premise in much of the DVFS work is that a system, a task, or a program can be slowed down with disproportionally small impact on its performance (or the perception of performance for interactive tasks), while at the same time obtaining significant savings in power consumption by voltage scaling. This can only be achieved by intelligently reducing frequency to remove *slack*: idle time in the system, slack in tasks with deadlines, or instruction slack due to memory accesses in memory-bound program phases. A similar idea can be applied at the hardware level. Ernst, et al. proposed a DVFS variation intended to remove slack in the timing of the hardware itself. Their approach is called Razor [73].

The driving motivation is to scale the supply voltage as low as possible for a given frequency while still maintaining correct operation. What prevents scaling the voltage below a critical level for a given frequency is the built-in margins in a process technology. For a given frequency, a voltage level is allowed which is safely above the lowest voltage level needed for the *worst-case process and environment variability* in the design. In other words, the relation between voltage and frequency is such that it guarantees correct operation with significant margin from the worst-case scenario.

This, however, diminishes the value of DVFS since the useful voltage range for DVFS shrinks with each new process technology. Going below the critical voltage level (*subcritical voltage*) for a given frequency invites trouble: timing faults. Faults, however, are unlikely to

---

[5]Similarly to the architecture described in Section 3.4.2 but allowing multiple cores to be active at the same time.

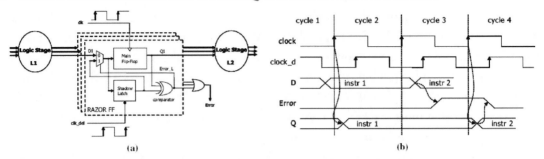

FIGURE 3.8: Razor flip-flop (on the left). Example operation (on the right). In cycle 2, incorrect results of the logic stage L1 are latched in the main latch. This is detected by the Razor flip-flop and the error is signaled in cycle 3. The error is corrected, the main latch receives the correct value from the shadow latch and operation is resumed in cycle 4. Reproduced from [73]. Copyright 2003 IEEE.

occur because worst-case conditions appear very infrequently. On the other hand, the more voltage is reduced below the critical level, the more likely faults become.

The idea behind Razor is to lower the voltage until *timing faults start to occur*. These timing faults are detected by the hardware itself using special, "safe" flip-flops that detect timing violations. Something analogous can be done by tuning the supply voltage during processor operation to meet a predetermined delay through an embedded inverter delay chain. This global approach still requires margins in the voltage levels, since it does not account for local variations. In contrast, Razor adapts dynamically to *local* conditions.

The *Razor flip-flop* (Figure 3.8) double-samples pipeline stage values: once with the fast clock and a second time with a time-delayed clock. This is done by using two latches. The main latch latches the data on the fast clock. A *shadow latch*, controlled by the time-delayed clock, latches the data for the second time. To guarantee that shadow latches always latch correct data, subcritical voltage is constrained at design time to levels that allow the shadow latch set-up times to exceed the delay of the logic—even in worst-case conditions.

The values latched by the main latch and the shadow latch are then compared with a metastability-tolerant comparator. If the values sampled with the fast clock are the same with those sampled with the delayed clock then no error has occurred. This is the common case. If, however, a timing error occurred then the values do not match. In this case, an error signal is raised by the faulting flip-flop.

The error signal restores the incorrect value of the main latch using the value of the shadow latch. All other Razor flip-flops in the same pipeline stage are also restored (regardless of whether they have generated a timing error). To recover the whole pipeline after a Razor error, two techniques can be used. The first uses clock gating to stall the whole pipeline for a single cycle so that the correct values can be restored in the flip-flops. Operation then resumes

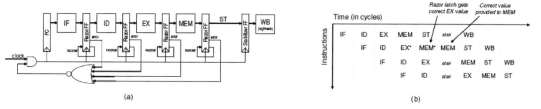

(a)   (b)

FIGURE 3.9: Pipeline recovery with global clock gating. A timing error in the EX stage results in the MEM stage receiving the correct value too late. Global clock gating stops the whole pipeline so the correct value can be restored in the Razor flip-flops. Operation then resumes one cycle later with the correct values. Reproduced from [73]. Copyright 2003 IEEE.

with the correct values. This guarantees forward progress since a faulting instruction simply continues execution in the next pipeline stage with the correct value. This technique is shown in Figure 3.9.

If global clock-gating is not possible in an aggressively clocked design due to impact on the cycle time, the second alternative technique: a *counterflow* pipelined approach—which scales well with clock cycle—can be used (Figure 3.10). In this case, the faulting Razor flip-flop distributes a *bubble* signal toward the end of the pipeline and a *flush* signal toward its front. The bubble ensures that the faulting instruction will take an additional cycle to complete its remaining stages, while the instructions following the faulting instruction are flushed.

Besides the timing errors that are detected by the Razor flip-flop, there is also a more serious error that can happen—albeit with a very small probability. This is when the Razor flip-flop detects that its error signal, after a timing error, is potentially in a meta-stable state. In this case, the Razor flip-flop raises a panic signal (two cycles after the actual timing error) to invoke higher-level control. Such an error is treated as a mis-speculation. A panic signal from the faulting Razor flip-flop flushes the whole pipeline, including the faulting instruction, restores the correct state, and replays the affected instructions.

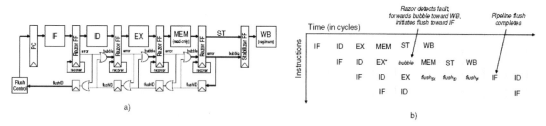

a)   b)

FIGURE 3.10: Counterflow recovery of the pipeline after a timing fault. *Flush* and *bubble* signals are distributed from the faulting Razor flip-flop toward the the front- and the back-end of the pipeline, respectively. Reproduced from [73]. Copyright 2003 IEEE.

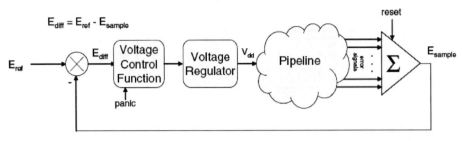

FIGURE 3.11: Razor voltage control system. The error rate is measured (E_{sample}), compared to a reference error rate (E_{ref}) and a voltage level is selected to minimize their difference. Reproduced from [73]. Copyright 2003 IEEE.

Timing errors, and the more rare meta-stability errors, have a negative impact on performance. Absence of errors, on the other hand, could mean that voltage is not scaled aggressively enough for the frequency. The Razor voltage control system is based on monitoring the error rate and works to maintain a constant error rate, E_{ref} If the measured error rate E_{sample} is less than the reference rate, then the voltage is further scaled downwards. In the opposite case, the voltage is increased to reduce the error rate.

Razor has been tested in real chips (modified ARM cores) and has been shown to work well. Substantial power savings (64% reported in their first paper [73]) can be expected by breaching the voltage margins in the designs. But lowering the voltage for any significant gain is also bound to cause a significant number of timing errors which will then affect both performance and energy. Performance overhead from timing misspeculations is at about 3% in their first study.

CHAPTER 4

Optimizing Capacitance and Switching Activity to Reduce Dynamic Power

The capacitance (C) and the switching activity factor (A) in the dynamic power equation are intimately intertwined. In fact, it is not uncommon to see the dynamic power equation written as: $P = V^2 \times f \times C_{\text{effective}}$, where $C_{\text{effective}}$, the *effective switched capacitance*, is the product of A and C. The switching activity factor, a fraction between 0 and 1, simply expresses how much of the circuit's *total* capacitance is charged and discharged—on average—during each cycle at the operating frequency f.

Undoubtedly, architecture and microarchitecture exert fundamental influence on both capacitance and switching activity. It is generally accepted that instruction-level parallelism (ILP), or more precicely the complexity required to *dynamically* uncover and exploit ILP, brought dramatic increases in both factors. This, coupled with the phenomenal frequency increase in CMOS technology, resulted in the exacerbated power problem of today. The complexity of a processor, how aggressively it exploits parallelism—in particular ILP—the bit-width of its structures, etc., directly affect switching activity. On the other hand, the *size* of a processor's structures and how well it is organized to exploit locality (for instance, whether functional units are clustered or not) determine the number of transistors and their interconnects, hence directly affecting capacitance.

While capacitance is more effectively manipulated at the circuit and process technology level, switching activity is effectively optimized at the architectural or microarchitectural level. More importantly, while capacitance is determined and fixed at design time, switching activity can change by run-time optimizations. Not surprisingly, a plethora of architectural techniques focus on dynamically optimizing switching activity.

Chapter structure: Fundamentally, the majority of the proposed techniques aim at eliminating unnecessary switching activity during computation; we call such switching *excess switching activity*. The trick is to find it and eliminate it, without inordinately harming performance. In many cases, such techniques inadvertently affect capacitance—at design time. Because they can affect both factors, more broadly, their goal is to reduce the *effective switched capacitance*. An orthogonal approach to such optimizations is to *dynamically steer work* to structures designed to offer a range of power/performance levels.

The structure of this chapter is based on a categorization of the excess switching activity (and by extension a categorization of the effective switched capacitance). The following section, Section 4.1, presents this categorization (summarized in Table 4.1) and creates the road map for the main body of the chapter. Work steering approaches conclude the chapter (Section 4.13).

4.1   A ROAD MAP FOR EFFECTIVE SWITCHED CAPACITANCE

Techniques to reduce effective switching capacitance have been developed for every major processor structure—the datapath, the ALU, the caches, the instruction scheduling hardware, and so on. In many cases, techniques for different structures share a common line of attack because they focus on the same root cause of *excess switching activity*. Motivated by this observation, the main body of this chapter unfolds based on a categorization of the excess switching activity (presented in Section 4.1.1).

The categorization of excess switching activity also implies a corresponding categorization of the *effective switched capacitance* when one takes into account the potential impact on capacitance. Section 4.1.2 gives a short overview on how architectural techniques presented in this chapter can affect capacitance. However, for the rest of the chapter, we will not explicitly refer to capacitance except in a few places where it is needed.

Finally, the chapter concludes with a section on *dynamic work steering* (Section 4.13). Approaches in this class statically provide multiple units that can perform the same work but with different power/performance characteristics. Rather than trying to optimize away excess switching activity in a single unit, these approaches simply steer computation to the appropriate unit according to runtime requirements for power or performance. Because any of the types of excess switching activity can potentially be handled with dynamic work steering, we consider this strategy to be orthogonal to the optimization techniques for excess switching activity.

4.1.1   Excess Switching Activity

To classify excess switching activity, we ask the following question: why is it there in the first place? We distinguish a number of causes, which lead to seven different activity types. These types are summarized in Table 4.1, along with the corresponding causes for their existence (column two), the typical granularity where they appear (column three), a quick

TABLE 4.1: Types of switching activities that can be reduced to save power. An additional *Line of Attack* that can potentially apply to many types of switching activity and different granularities is *Dynamic Work Steering*, presented separately in *Section 4.13*

Excess Switching Activity	Cause	Granularity	Line of Attack	Example Technique	Section
Idle-unit	Clock-induced switching in unused (idle) units	Functional unit	Clock gating	Clock-gated Functional Units [11, 218, 152, 57, 58]	4.2
Idle-width	Bit-width too wide for typical operations	Cross section of FUs, datapaths, caches	Adapting to narrow-width operands	Clock-gated high-order bits in ALUs [37, 44], cache compression [221, 235, 234, 237, 141]	4.3, 4.4
Idle-capacity	Processor structures sized to support peak ILP not fully utilized in typical programs	Large processor structures: instruction queues, core width, caches	Dynamic resizing	instruction queue resizing [42, 80, 182], cache resizing [244, 8, 21, 68, 9, 168, 241, 131]	4.5, 4.6, 4.7, 4.8
Parallel-speculative	Parallel (speculative) activity for speed	Caches, coherence H/W	Serializing or filtering parallel activity	Way prediction and other techniques for set-associative caches [95, 87, 133, 109, 183, 242, 249, 168, 241, 131] Coherence, [171]	4.9
Cacheable (repetitive)	Repetitive computing with the same inputs, or repetitive memory accessing	Architectural structures: FUs, caches	Caching—or memoization	Work reuse [56, 86, 107, 208], filter cache [142], loop buffers [150, 24, 25, 232, 10, 110], trace caches [193, 210]	4.10
Speculative	Activity wasted on wrong speculation	Out-of-order core	Execution throttling	Pipeline & selective gating [161, 16]	4.11
Value-dependent	Data value encoding not optimal	FU, datapaths	Applying different data encoding	Bus encodings [176, 75, 27, 28, 173, 212, 26, 188, 55, 233]	4.12

description of the prevailing line of attack (column four), and example techniques used to eliminate this type of switching activity (column five). Here, we describe each of the seven activity types in more detail.

- *Idle-Unit Switching Activity*: This is excess switching activity triggered by clock transitions in otherwise unused portions of the hardware. The classic example of this type is an ALU adder that switches every single clock cycle even when it is not producing any useful results. The straightforward solution is to gate the clock to the whole unit (Section 4.2).

- *Idle-width switching activity*[1]: This type of activity is the result of a mismatch in the implemented width of various processor structures (datapaths, ALUs, register files, caches) and the actual width used in many common operations. For example, operating on 8-bit quantities in 32-bit hardware entails unnecessary switching simply because the original design is wider than what is actually needed for such operations. The solution is to dynamically *adjust the hardware to narrow width operands*. Techniques dealing with this type of activity encompass datapaths, sequential logic (Section 4.3), and caches (Section 4.4). Value compression also falls in this category, considering compressed values as a special case of narrow-width values (since all the significant information is carried with fewer bits).

- *Idle-capacity switching activity*[1]: Another type of "over-provisioned" activity results when a program does not use the provided hardware structures in their entirety. This is not to say that the original design is over-provisioned in all cases. In fact, processor structures are finely (and laboriously) balanced to provide peak performance for target workloads. Rather, the wording is meant to stress that sometimes there is a mismatch between what is offered and what is needed and this creates opportunities for dynamic capacity optimizations. Idle-capacity switching activity is especially evident in large CAM structures (e.g., instruction queues), where searching entails activity proportional to size. Again, this type of activity can be avoided by *resizing structures* to match program behavior (Section 4.5). Often, dynamic resizing also affects capacitance requiring the segmentation of long wires into smaller segments. Relevant techniques for reducing the switching activity of the instruction window hardware, the core (as a whole) and the caches are presented in Sections 4.6, 4.7, and 4.8, respectively.

---

[1] *Idle-unit*, and *idle-width/idle-capacity* are similar to some degree—useless activity that can be eliminated—but a distinction is made so that the first (*unit*) concerns excess activity in whole units or structures and the last two (*width* and *capacity*) concern excess activity in *slices* across units or structures. The distinction is subtle and has to do with program behavior: *idle-unit* switching activity has to do with what resources the program *does not use in their entirety*; *idle-width* or *idle-capacity* switching has to do with *partial use* of resources.

- *Parallel switching activity*: Significant activity is also expended in parallel for performance. An example is the parallel search of all the ways of an associative cache while it is known beforehand that all but one—at most—will fail to produce a hit (Section 4.9). Another example is the parallel activity in snoopy caches that keeps them coherent. Some of this parallel switching activity *can be eliminated at the expense of some performance*. In contrast to *unnecessary* or *over-provisioned* activity which, theoretically, can be eliminated without an impact on performance, reducing parallel switching activity requires sophisticated schemes that trade performance for power (Section 4.9).

- Cacheable switching activity: Repetitive switching activity can be avoided if it relates to operations whose results can be cached and reused. The repetitive nature of this switching activity is due to program locality. Once we recognize such behavior, we can *convert computing activity to cache lookups*, which are designed to use less power on average (Section 4.10). A real-world example are the trace caches in the Intel P6 architecture, which are intended to avoid expensive decoding/uop translation of frequently reappearing IA-32 instructions. Caching cache operations (reads and writes) lead to techniques such as filter caches or loop buffers (Section 4.10).

- *Speculative switching activity*: This type of activity is unique to out-of-order processors supporting speculative execution. Speculatively executing incorrect instructions is wasted activity. The line of attack here is to *throttle speculative execution* when the confidence for its usefulness is low (Section 4.11).

- *Value-dependent switching activity*: Power consumed in this case depends on the actual data values. Thus, a different encoding of the data can potentially reduce power consumption. Although switching activity in circuits largely depends on their inputs (e.g., the switching activity of an adder depends on what it adds), one of the most successful areas for applying data encodings is data communication. The prime example is bus switching activity where power is consumed only if bus lines are switched to different logic values (Section 4.12). In this case, a *different data encoding* can reduce bus line transitions.

4.1.2 Capacitance

The total capacitance of a chip, C, does not change dynamically; it is fixed at design time. Architecture, microarchitecture, and circuit design, however, do influence the magnitude of C by dictating both the total number of transistors and their interconnect. In an implementation, floorplanning and place-and-route methodologies also play a significant role in determining C, affecting the actual length of the wires on chip, but we do not examine them here. We limit our discussions on how architecture affects C—for example: via the size of structures, via locality, or by breaking up larger monolithic structures into smaller chunks.

While the total capacitance of a chip depends on both the capacitance of its transistors and the capacitance of its wires, the latter becomes increasingly important with every process generation. The dynamic power formula actually describes the power expended by charging and discharging the node capacitance at the output of every logic gate. Burd and Brodersen [39] describe this capacitance, C_L, as: $C_L = C_W + C_{Fixed}$.

The first term, C_W, is the product of a technology constant and the device width, W. For a single node, C_W consists of the input capacitance of the subsequent gates plus some of the diffusion capacitance of the node's output. Although there is a significant control over the device width W, its optimization becomes a complex interplay between power and delay which is best handled at the cicruit level.

Architecture, on the other hand, largely determines the number of nodes required in a design (e.g., structure sizes) and their fan-out. Some low-power architectural techniques aim to reduce the total effective capacitance by *adding more* hardware which raises the total capacitance. The goal in this case is to reduce the overall switching activity enough to produce net gains in power consumption.

The true playground for the architect is the second term, C_{Fixed}, which is composed of the remaining part of the diffusion capacitance of the gate's output (not dependent on W) and the capacitance of the (wire) interconnect connecting the gate's output to the inputs of subsequent gates. Wire capacitance can be reduced by effective placement and routing, but also by architectural choices. More importantly, reducing wire capacitance yields benefits simultaneously in power and speed (latency).

Wire capacitance directly affects lanetcy which is determined by the wire's $R \times C$ product. Architectural techniques to aleviate latency (for example, clustered functional unit organizations, multiscalar-like architectures [209], NUCA caches [140], tiled CMPs, etc.) also affect C. Consider, for instance, that a non-uniform cache architecture (NUCA) breaks up large monolithic memory banks into smaller chunks to address long wire latencies found in bit-lines and wordlines. In the process, however, this changes the effective capacitance of the cache.

Besides large-scale architect ural choices that enhance locality (i.e., promote the use of shorter wires), many low-power architetcural techniques require the partitioning of large structures into smaller segments. Such techniques are commonly based on two circuit-level techniques that affect wire-capacitance: *wire partitioning* and *bit-line segmentation*.

- *Wire partitioning*: This technique breaks up long wires in order to reduce their wire delay. The resultant segmented wire can be shortened by electrically "turning off" part of it. This eliminates switching in the disabled part, and reduces the capacitance of the active part (see "Sidebar: Wire Partitioning").

- *Bit-line segmentation*: This technique reduces the apparent capacitance of a long bit-line, as seen by the memory cells and the peripheral circuitry (precharge circuits and sense amps). The net effect is that for the same cache activity, less capacitance is switched. This technique is further described in "Sidebar: Bit-line Segmentation".

4.2   IDLE-UNIT SWITCHING ACTIVITY: CLOCK GATING

The techniques presented in this section aim to reduce or eliminate excess activity that does not have any effect on the computation being performed. This type of excess activity appears at different granularities: from the tiniest circuits and individual flip-flops, to whole functional units, or even larger structures and whole subsystems (e.g., memory, I/O, CPU). Idle-unit switching activity is caused by the clock being fed to an *idle unit*—at any of the granularities mentioned above—which, for that particular time, does nothing useful with respect to the computation being performed.

Clock gating: Gating the clock to the particular idle unit using a control signal is the way to eliminate such a switching activity. Clock gating is the central *mechanism* used in many of the techniques and policies that we discuss later in this chapter. However, we make a distinction between the mechanism of clock gating and the high-level policies that make use of this mechanism at various granularities. Higher-level policies are described according to the type of excess activity they were destined for. For instance, the following two *applications* of clock gating are not simply lumped under the current heading but are explained elsewhere.

- *Value gating*, which is presented in Section 4.3, has to do with the unused *width* in the datapath in the presence of *narrow-width* operands.
- *Pipeline balancing*, which is presented in Section 4.5, is a way to adjust the issue width of an out-of-order microarchitecture to fit program needs.

In both of these examples, it is the high-level policy which engages clock gating that is of interest and not the clock gating mechanism *per se*. In this section, we concentrate solely on the mechanism of clock gating as it applies to basic circuits, to larger logic blocks, and finally to the processor core.

4.2.1   Circuit-Level Basics

At the granularity of a small circuit or an individual flip-flop, clock gating reduces power by preventing unnecessary charging and discharging of the circuit's capacitances [152]. For a flip-flop, the capacitance of interest is the cumulative capacitance connected to the clock. This capacitance, shown as C_g in Figure 4.1, is charged and discharged in every clock cycle. Using an AND gate to gate the clock with a control signal, we replace the capacitance of the flip-flop with the capacitance of the AND gate when the control signal is 0: the AND gate transistors

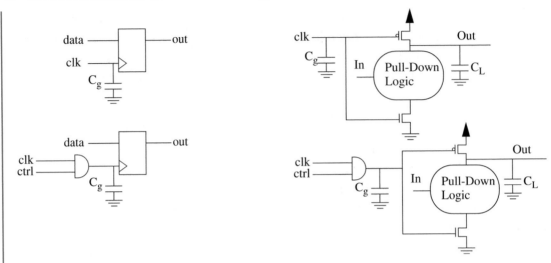

FIGURE 4.1: Clock gating in flip-flops and dynamic cells.

switch with the clock, but the flip-flop transistors do not. Since the capacitance of the gate is much smaller than that of the flip-flop, power is saved.

For logic we distinguish two cases: static and dynamic (domino) logic. For static logic, to eliminate any switching it is enough to prevent the inputs from changing. Usually, this can be accomplished by clock gating the input latches (or flip-flops). However, in the case of dynamic logic, which constitutes the bulk of the high-performance logic in a processor, things are different. Power *can* be consumed even if the inputs to the circuit *do not change*. In dynamic logic, the output is precharged to V_{dd} and a pull-down network can discharge the output node to Gnd if needed. So, if the logic function implemented by the circuit evaluates to 0, its output node will be charged and discharged in every cycle, even with constant inputs. Thus, in a dynamic logic cell, besides the capacitance connected to the clock (C_g), its *own* output capacitive load (C_L) comes into play and can switched (precharged and discharged) every cycle.

At this granularity, clock-gating can be applied at the Register Transfer Level (RTL) with simple transformations on the unoptimized designs. For example, in flip-flops where the input is guarded by a condition, this condition can be used to gate the clock instead (Figure 4.2, left side). When such a flip-flop is part of a pipeline, the same condition can be propagated to gate subsequent stages (Figure 4.2, right side) using additional latches that are always clocked. Because additional hardware is required for clock gating, the benefit should be large enough to warrant it. Today, such transformations are routinely applied by all major RTL compilers. In contrast, at coarser granularities, high-level control policies are needed to decide when to gate the clock to whole blocks.

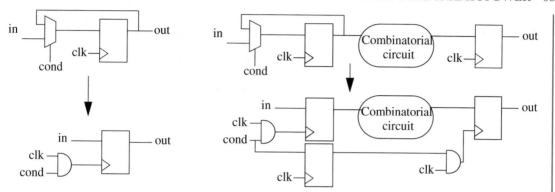

FIGURE 4.2: Automatic clock gating transformations at the RTL level.

4.2.2   Precomputation and Guarded Evaluation

At a coarser granularity, at the level of a logic block, systematic approaches for clock gating have been proposed. Two prominent techniques are *precomputation* [11] and *guarded evaluation* [218].

Introduced by Alidina, Monteiro, Devadas, Ghosh, and Papefthymiou, precomputation, as the name suggests, aims to derive a precomputation circuit for a logic block. The precomputation circuit under some condition subsumes the operation of the larger logic block. One example would be a *multiplexed* precomputation architecture in which a large logic block F is split into *two new separate blocks* $F(x = 0)$ and $F(x = 1)$, where x is a control variable that also drives the output multiplexor for these two blocks. Only the block producing the output needs to be evaluated, while the operation of the other block can be gated by x.

In contrast, *guarded evaluation*, proposed by Tiwari, Malik, and Ashar, aims to shut down—clock gate—*part of the original circuit* which, under some condition, is not needed in evaluating the final output. The idea (shown in Figure 4.3) is to find a subset F of the gates of the original circuit that generate a signal x. If there is a readily available signal S that determines whether the evaluation of the signal x has any effect on the final output of the whole circuit, then S can be used to clock gate F. In other words, S determines whether x is a *don't care* for the final output. If x is a don't care, F need not switch for the correct evaluation of the output. Even if S is not available at the start of the computation, it can still be used to advantage in stopping the evaluation of F. However, in such a case, partial evaluation of F takes away some of the power savings. Tiwari et al. develop a theoretical framework and the related algorithms for automatically determining parts of a circuit that can be disabled on a per-cycle basis. Applying their methodology on a number of benchmark circuits, the authors show substantial power savings.

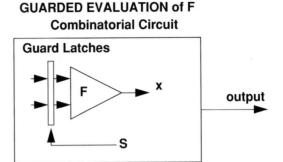

FIGURE 4.3: Guarded evaluation: a signal *S* determines when the output x of a set of gates F is a don't care for the generation of the final output. S can be used to gate the inputs to F, effectively shutting down F when it is not needed for the evaluation of the final output. Adapted from [218].

4.2.3   Deterministic Clock Gating

At a still higher level, at the level of processor core, deterministic clock gating—gating the clock to processor structures when they are known to be idle—provides noticeable power savings without performance loss. It safely improves EDP by not compromising performance [152]. With perfect clock gating, power savings (over the unoptimized design) are proportional to the average part of the hardware that a program leaves idle during each cycle of its execution.

Although the idea is straightforward and used in actual processors from early on [58], its application on a superscalar pipeline was not published until 2003. In the mean time, a few predictive clock-gating techniques had already been proposed (we will discuss some of those later in this chapter). Li, Bhunia, Chen, Vijaykumar, and Roy give a detailed description of deterministic clock gating in a superscalar pipeline [152]. They consider a high-performance implementation using dynamic domino logic for speed. This means that—besides latches—the pipeline stages themselves must be clock-gated. As per our discussion above, it is not enough to stop their inputs from changing; their clock must be stopped as well.

In both cases, the deterministic nature of clock gating stems from the ability to deduce a few cycles in advance the idleness (or use) of a latch or a pipeline stage. Once a decision to stop the clock is made, it is carried via additional (always clocked) latches to the correct stage or latch and delivered at the right cycle. Clock gating control proceeds down the pipeline similarly to a bubble.

Figure 4.4 shows the 8-stage, superscalar, pipeline used by Li et al. in their study [152]. The pipeline stages and latches that can be deterministically clock-gated are shown shaded. The decision to gate a stage (or the latches of a stage) must be taken at a previous pipeline stage—i.e., well in advance of the cycle that is gated.

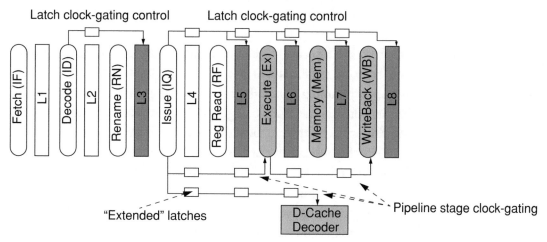

FIGURE 4.4: Deterministic Clock Gating. Pipeline latches and pipeline stages that can be clock-gated are shown shaded. Adapted from [152].

The Fetch and Decode stages and their latches are never clock-gated since instructions are needed almost every cycle. Decode stage information (i.e., number of decoded instructions) determines the glock gating of the issue latches (L3).[2] Unfortunately, there is no time to gate the L2 latches by the time this information is available. Although the Rename stage could be gated, it is also left clocked.

Decode stage information can be used to clock-gate up to the Issue stage. There, instructions are buffered in an instruction queue. Further power optimizations for the instruction queue are presented in Section 4.5. Every cycle, a decision is made anew by the scheduler on which instructions to forward to the next stage for execution. This information, passed via the "extended" always-clocked latches, can now be used to clock-gate all the subsequent pipeline latches, as well as the execution units, the data cache—in particular its ports—and the result-bus drivers.

Issue-stage information comes too late to clock-gate the Issue-latches themselves (L4) or the Register Read stage, but on time to clock-gate the rest of the latches (L5, L6, L7, and L8). Similarly, execution units can be clock-gated with Issue-stage information. At issue time, we know what functional unit is needed and by when.

Load/store instructions determine the clock gating for the data cache. Depending on the number of memory accesses during a cycle, data cache ports—essentially the dynamic logic of the decoder and its wordline drivers—are clock gated. Although this is a straightforward

---

[2]A similar concept but at a finer granularity is used to clock gate the unused width of the datapath; this concept is presented separately in Section 4.3.

application of clock gating, more sophisticated techniques to reduce cache power are presented in Sections 4.4 and 4.9.

Finally, information from the Execute stage is used two cycles later to clock-gate the Writeback stage. In the Writeback stage, data is put on the result bus and routed to the Instruction queue to wake up any waiting instructions. Clock-gating in the Writeback stage is somewhat different than in other dynamic-logic stages. Power is consumed only when bus lines switch logic levels. Additional techniques to reduce bus switching are presented in Section 4.12. To prevent spurious switching when the result bus is idle, the data latches feeding the bus-line drivers are clock-gated to shield bus lines from any changes.

Li et al. evaluate deterministic clock gating (DCG) [152] with Wattch. By applying DCG to all the latches and stages described above, they report power savings of 21% and 19% (on average) for the SPEC2000 integer and floating point benchmarks, respectively. They also compare deterministic clock gating to a predictive clock gating technique, *Pipeline Balancing (PB)* [19], which is presented in detail in Section 4.7. Pipeline Balancing adjusts the width of the superscalar pipeline by gating functional unit clusters to match the needs of programs. Being a *coarser grain* technique, it misses some of the opportunities to gate idle hardware, resulting— according to Li et al.—in power savings of less than 10%. PB, being also a *predictive* technique, can also err and negatively affect performance (incurring a 2–3% slowdown on average) which is why DCG fares even better (compared to Pipeline Balancing) in terms of EDP.

4.2.4   Clock gating examples

Today, virtually all processor designs use clock gating to some degree. It is interesting that not only low-power designs but also many high-performance processors utilize extensive clock gating because of its nonextant impact on performance. Two prominent examples (one high-performance, the other low-power) are the IBM's Power5 [57] and Intel's XScale processors [58].

Power5: Dynamic clock gating is extensively used in this high-performance IBM processor [57]. According to IBM, the use of clock gating yields a reduction in switching power by more than 25% without affecting either performance or frequency. The larger the unit that is clock gated, the more likely it is to cause di/dt problems, i.e., large swings in the current of the power rails that charge these units when the clock signal is reinstated. For this reason POWER5 implements *fine-grain gating domains*, limiting the induced noise. All clock gating events are programmable, which allows extensive control over clock gating. Figure 4.5 shows a clock gating circuit (adapted from [57]) of the Power5. There are both global and local clock gating *enable* signals but the actual gating decision is taken by logic dedicated to each gated unit. This logic produces the *Dynamic Stop* signal depending on the usage characteristics of the gated unit.

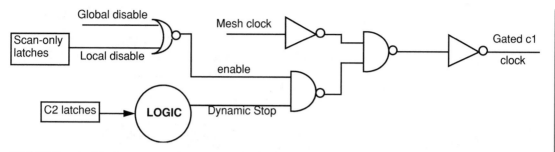

FIGURE 4.5: Clock gating in the Power5. Adapted from [57].

XScale core: The Intel Xscale processor is a low-power processor boasting an impressive DVFS range. While its low power abilities come mainly from DVFS, it is also highly optimized for low power at the circuit and architectural levels [58]. The design of the processor is mostly static CMOS logic supporting full clock stop. The processor implements three power-saving modes (besides its extensive DVFS abilities): *Idle mode* (full clock stop via clock gating), *Standby mode* which stops the phased locked loop (PLL) and puts the processor in reverse body bias for low leakage, and finally *Sleep mode*, which does not even retain state.

The low-power design features of an Xscale core are detailed by Clark et al. [58]. At the circuit design level, the implementation utilizes *pulse-clocked latches* instead of ordinary master–slave latches, cutting down on clock power consumption by 30%. Pulse-clock latches do not need explicit clock gating (as was described in Section 4.2.1 for ordinary flip-flops) and result in less switching activity for the sequential elements they feed.

Clock pulses to drive the pulse-clocked latches are generated by distributed units called Local Clock Buffers (LCBs), which are fed by a balanced global clock network. Clock gating in XScale is implemented at the LCBs. Each LCB has enable signals which can stop the pulse generation. Because of the overhead of pulse generation and clock gating, each LCB must feed at least five latches. This is the smallest unit in the Xscale core that can be individually clock gated.

Clock gating in the Xscale is implemented at three different levels. First, at the PLL to implement the processor-wide Idle mode by halting all clock activity; second, at the global clock level (GCLK) with 83 unique enable signals; third at the individual LCB level with 400 distinct enable signals [58]. Although no further details are disclosed for the policies to engage clock gating, deterministic clock gating described above can be easily implemented in such a framework.

Xscale cache: Architecturally, the Xscale is a simple, single-pipeline, in-order processor. The pipeline is 7-stage for integer operations, 8-stage for memory, and 9-stage when executing the compact ARM Thumb instruction set. The pipeline is optimized for the high-frequency operation and its low complexity makes it very power efficient.

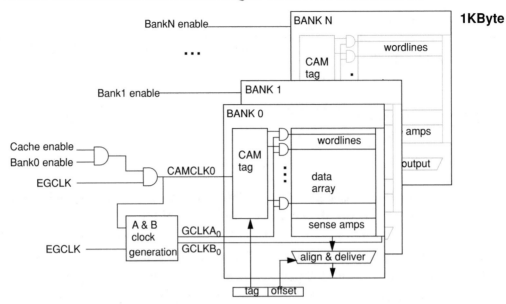

FIGURE 4.6: 32-Bank CAM-tag cache in Xscale. Adapted from [58].

One of Xscale's distinguishing architectural features for low power is its CAM-tag organization of its 32KB instruction and data caches. A CAM-tag cache organization (as opposed to a RAM-tag organization) combines address decoding with tag comparison in one step. It allows highly associative caches (e.g., 32-way in the Xscale) with very low miss rates while, at the same time, being very power efficient at that performance level [244].

Figure 4.6 shows the organization of a 32KB cache in the Xscale. The cache comprises 32 independent banks of 1KB each. Each bank is composed of a CAM-tag array and a data array. A tag match in the CAM drives the corresponding wordline of the data array. The cache is extensively clock-gated: only one of the 32 banks (1KB) is enabled during an access. This limits the CAM rows that are searched to the rows of a single 32-way set. Once the clock for the CAM tag match of a bank is gated (the CAMCLK in Figure 4.6), no additional clocks can be generated for that bank (i.e., clocks A and B in Figure 4.6) inhibiting any further activity. Clark et al. emphasize that this extensive clock gating goes beyond any previous design in making this a very power-efficient cache [58].

4.3   IDLE-WIDTH SWITCHING ACTIVITY: CORE

Idle-Width switching activity is the excessive switching activity which arises from a mismatch between the designed bit-width of a processor and the actual bit-width needed in frequently occurring operations. It is specific to operations, i.e., it appears when operations are performed—whether ALU or memory operations (e.g., cache reads and writes). The main approach to address this type of activity is to dynamically detect *narrow-width* operands and to either adapt the width of the machine accordingly or, alternatively, pack multiple narrow-width operations together. This goes for both the datapath, which we discuss here, and the caches discussed in (Section 4.4).

The meaning of "narrow-width" has changed over time. Starting from a simple definition—a value having only few significant bits in the low-order bit positions—the term evolved to *significance compression* by relaxing some of the initial constraints; full-fledged *frequent value compression*, that creates "narrow values" in caches, is a small step further.

4.3.1   Narrow-Width Operands

One of the most prominent characteristics of a processor is the data width for which it is designed and built. While there are plenty of older-generation low-power, low-performance, 8-bit processors, most recent designs are either 32- or 64-bits wide. Quite often, however, this width is not exercised in operations. For instance, most address offsets are small and can easily fit in 16 or even 8 bits [96], so address arithmetic does not use all of the provisioned width of the machine. The difference between the provisioned width and the width of commonly used operands is increased with the move to 64-bit architectures where most of the width of the datapath remains unused.

Multimedia applications also use many small operands. In fact, this has been the motivation behind SIMD ISA extensions such as Intel's MMX and SSE/2/3, IBM's AltiVec, and others [187, 66]. These multimedia ISA extensions pack multiple sub-word operands in the width of the machine and operate on all of them in parallel.

Brooks and Martonosi were among the first to propose techniques addressing Idle-Width switching activity [37]. The motivation in their work is the abundance of narrow-width operands in integer and multimedia applications. Figure 4.7 reproduces some of the relevant statistics from their work for applications in the SPECint95 and the MediaBench benchmark suites. The statistics show operand widths in the 64-bit Alpha ISA (with the DEC *cc* compiler using aggressive optimizations). The left graphs show that a respectable portion (50%) of the operations have both their operands "narrow" (16-bit wide or less). Furthermore, very few operations have operands wider than 33 bits in a 64-bit machine (middle graph).

Exploiting narrow-width operands can be done in one of two ways:

1. By disabling the unused width of the hardware and eliminating switching in the hardware *slice* that does not carry significant bits.

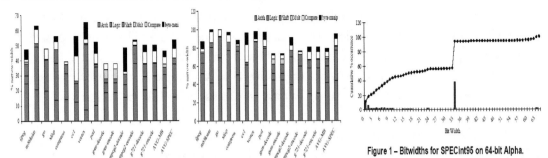

Figure 1 – Bitwidths for SPECint95 on 64-bit Alpha.

FIGURE 4.7: The width of operands in SPECint95 for the 64-bit Alpha ISA From [37], Copyright 1999 IEEE.

2. By packing more than one narrow-width operation in the full width of the hardware. Although this does not reduce power, it improves the performance and energy per operation, resulting in better EDP.

Although these two techniques can share a common infrastructure in determining narrow-width operands, each serves a different purpose—the first directly reduces power while the second aims to increase performance. Both can co-exist in an implementation and either one can be activated at any given time according to system-wide needs. There is also the possibility of using both techniques concurrently, packing operations whenever possible but disabling high-order bits of the ALU when there is only a single (narrow-width) operation passing through.

Dynamically detecting narrow-width operands: Data in Figure 4.7 show that applications have operands whose significant bits span the full range of bit widths.[3] So where is the line drawn between narrow and wide? Initially, a fixed division at 16 bits was proposed: anything 16-bit wide or less is defined as narrow width and is treated as a 16-bit quantity; anything greater than that is taken to be a full 64-bit operand [37]. As Figure 4.7 attests, this division defines about one-half of the operands as narrow.

The detection of narrow-width operands is accomplished dynamically (although static compiler analysis can also be a useful tool). Every value created in the ALU or loaded from the cache is checked for its effective size. If 48 of its leading bits are all zeros or ones (for a two's complement negative value), it can be represented and manipulated with just its lower 16 bits. In this case, the value is tagged as narrow. A single bit "width" tag follows the value throughout the machine.

Value gating: disabling the unused width: This technique directly reduces power by eliminating useless switching. This is done by disabling switching in the unused parts of the ALU

---

[3]Note that, for now, we assume that significant bits are strictly low-order bits. Subsequently, we relax this restriction.

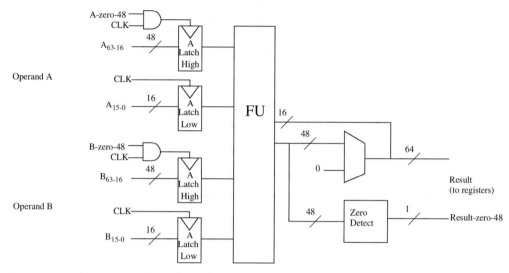

FIGURE 4.8: Clock-gating ALUs for narrow-width operands. Adapted from [37].

if both operands of an operation are tagged as narrow [37]. Potentially, part of the ALU could be safely disabled even if only one of the arguments is narrow but it would be too complicated to guarantee this beforehand. If both arguments are narrow, only their 16 lower bits are latched on the ALU latches. This ensures that no switching occurs for the 48 upper bits and is akin to clock gating the unused portion of the ALU (for static logic). To create the correct wide result, the appropriate bits (zeros or ones) are multiplexed on to the result bus. This technique yields significant power savings for the CPU's integer unit comprising of an adder, a booth multiplier, bit-wise logic, and a shifter. Specifically, in an Alpha-class, 4-instruction-wide superscalar, the average power consumption of the integer units can be reduced by 55% and 58% for the SPECint95 and the Mediabench benchmark suites, respectively [37].

Operation packing: packing narrow-width values: While Value Gating adjusts the width of the machine to the operand width, in this technique two operations with narrow-width operands are simultaneously issued to a single full-width ALU. This can increase performance (if there is contention for the ALUs) without incurring significant power overhead (since switching activity remains approximately the same) and as a consequence improve EDP.

The implementation is simple: issue logic detects two instructions that perform identical operations, are ready-to-issue, and that have all of their operands tagged as narrow-width. A set of multiplexors shifts the significant part of the operands of one of the operations into the high-order bits of the ALU inputs. The significant parts of the operands of the other operation remain at their normal position in the low-order bits. The combined operations are executed in the ALU in SIMD mode, similarly to SIMD multimedia extension instructions. The CPU is assumed to have support for SIMD instructions, e.g., it can segment the carry chain at 16-bit intervals. The overhead for this technique consists of the set of multiplexors to shift operands into high-order bits and the result to low order bits. Additionally, increased complexity is required in the issue logic.

Speculative-operation packing: Packing operands together effectively increases the CPU's issue bandwidth but the resulting speedup is small. The problem is that both operands of an operation are required to be narrow-width in order for packing to be considered. This is too restrictive. The odds of packing operations together can be significantly improved if we simply require at least one operand to be narrow width, but not necessarily both. There is a good chance that an operation between a narrow and a wide operand will not affect high-order bits. In this case, the high-order bits of the wide operand can be carried over to the result, while its low-order bits form a narrow operand.

It would be, however, too complex to guarantee, before performing the actual operation, that the high-order bits remain intact. The solution is provided by architectures supporting speculative execution: operations can be packed speculatively, as if their arguments were all narrow-width. If something goes wrong, the packed operations are squashed and re-executed (replayed) separately. The telltale sign of something going wrong is an overflow in the segmented carry chain, meaning that high-order bits are indeed affected by the operation.

This optimization brings the speedup of packing narrow width operations to approximately 4% for SPECint95 and 8% for Mediabench for an Alpha-class, 4-instruction-wide, superscalar CPU [37]. Speedup increases with the width of the machine as more instructions become available to choose from and pack together. This is a good result considering that the 55–58% power savings mentioned above for the first clock gating technique concern the integer units only, which in reality consume about a 10% of the processor's total power.

4.3.2   Significance Compression

Until now we have discussed a fixed-width definition (16-bits) of narrow-width operands and allowed for significant bits only in low-order positions. Relaxing these two constraints leads to a more general approach proposed by Canal and González [44], called *significance compression*. The idea is to *compress* non-significant bits (strings of zeros or ones) *anywhere* they appear in the full width of an operand. Each 32-bit word is augmented with a 3-bit tag describing the "significance" of each of its four bytes. A byte can be either *significant* or a sign extension of its preceding byte (i.e., just a string of zeros or ones). Of course, the first low-order byte cannot be a sign extension of any other byte and is always taken to be significant. The tags encode the significance of a byte in the manner shown in Table 4.2.

Canal and González report that the majority of values (87%) in SPECint and Mediabench benchmarks can be compressed with significance compression [44], although a good 75% of

TABLE 4.2: Byte Significance in a 32-bit Word. Adapted from [44].

Tag (Arbitrary Enumeration)	Byte Significance (S) (e for sign extension)	% of values in SPECInt95 and Mediabench
0	eeeS	61
1	eeSS	14
2	SSSS: NON-COMPRESSED	13
3	eSSS	7
4	SSeS	2
5	SeSS	2
6	eSeS	1
7	SeeS	1

all values is narrow-width using the Brooks–Martonosi 16-bit definition (i.e., only the first and possibly second bytes are significant). In other words, relaxing the two restrictions of the original narrow-width definition gives an additional 12% of "compressable" values.

Byte-serial pipelines: A complete pipeline can be built around this significance compression scheme [44]. Only significant bytes flow through the pipeline and are operated on. The rest are carried and stored via their tags. This opens up the possibility of a very low-power *byte-serial pipeline*, which is a single byte wide. If more than one significant byte needs to be processed at a pipeline stage, then this stage simply repeats for the significant bytes. However, the performance hit on this pipeline is severe: CPI (cycles per instruction) increases 79% over a full width (32-bit) pipeline. Activity savings range from 30% to 40% for the various pipeline stages.

A significant improvement in performance, without necessarily increasing switching activity, can be gained by doubling the width of the pipeline from one to two bytes in selected stages. This is done for the register file read and in the ALU stages which are the major bottlenecks. In this *byte semi-parallel pipeline*, the additional byte is enabled if there is more than one significant byte to be processed. This results in a performance hit over a full-blown pipeline of just 24% while retaining the 30–40% activity savings.

Increasing the pipeline width to four bytes (*byte-parallel pipeline*) and enabling only the parts that correspond to the significant bytes of a word, retains most of the activity savings and further improves performance, bringing it very close (6–2% slowdown depending on optimizations) to a full pipeline operating on uncompressed operands [44]. The byte-parallel pipeline brings us back to the first narrow-width technique which gates unused high-order bits albeit operating at a different granularity (at the byte level) and without requiring that significant bits be consecutive LSBs.

4.3.3 Further Reading on Narrow Width Operands

The idea of narrow width values has been applied to other structures as well. Ergin, Balkan, Ghose, and Ponomarev apply it to register files [72]. The intent is not so much to reduce power consumption, but to alleviate register pressure by making better use of the available physical registers. Similarly to packing two narrow values in the inputs of functional units or packing compressed lines in caches, multiple narrow values are packed in registers.

A number of these values can be packed in a register either "conservatively" or "speculatively." Conservatively means that a value is packed only after it is classified as narrow. This happens after a value is created by a functional unit. When a narrow value is packed in a different register than the one it was destined for, the register mapping for the packed value is updated in all the in-flight instructions. In contrast, "speculative packing" takes place in the register renaming stage, without certain knowledge of the width of the packed value. Packing and physical register assignment is performed by predicting the output width of instructions. The prediction history (per instruction) is kept in the instruction cache. The technique works well for performance—increases IPC by 15% in the SPEC2000—but may not offer significant advantages for power due to its complexity.

A different approach is followed in the work of Rochecouste, Pokam, and Seznec [192]. What they propose is to design a processor with dedicated narrow width datapaths—a width-partitioned microarchitecture (WPM). This is a *work steering* technique for this type of excess activity and is detailed in Section 4.13.

Finally, a scheme to pack multiple compressed instructions to improve instruction fetch bandwidth and power has been proposed by Hines, Green, Tyson, and Whalley [100]. But because this scheme uses Frequent Value Compression, which is explained next, we leave the details for the end of Section 4.4.

4.4 IDLE-WIDTH SWITCHING ACTIVITY: CACHES

Techniques addressing idle-width activity can be also extended to cache operations (reading and writing the cache). For example, power can be saved by accessing only the significant or the *compressed* part of a word. This results in reading or writing *fewer* bits, and corresponds to clock gating unused parts of the ALU or the datapath. Alternatively, multiple cache lines can be compressed and packed in the space of an uncompressed line. This improves the performance

TABLE 4.3: Cache Compression Techniques

Technique	Compression	Power Benefit
Dynamic Zero Compression, Villa, Zhang, and Asanović [221]	Limited significance compression (only zero bytes)	Accessing only significant bits
Frequent Value Cache (FVC), Yang and Gupta [235, 234]	Frequent value compression	Accessing a small index to a dictionary containing the value instead of the full value
Compression Cache (CC) Yang, Zhang, and Gupta [237]	Frequent value compression	Accessing a small index to a dictionary containing the value instead of the full value, Increasing efficiency (hit rate) by packing more lines in the cache
Significance compression cache, Kim, Austin, and Mudge [141]	Significance compression	Accessing only significant bits, Increasing efficiency (hit rate) by packing more lines in the cache

of the cache and corresponds to the dynamic packing of narrow-width operands in the datapath [37]. Table 4.3 lists the techniques presented in this section.

4.4.1   Dynamic Zero Compression: Accessing Only Significant Bits

A representative technique for accessing only significant bits is *Dynamic Zero Compression* (*DZC*) proposed by Villa, Zhang, and Asanović [221]. This technique uses significance compression, anywhere in a word, at a byte granularity, but it only compresses zero bytes (not string-of-ones bytes). Zero bytes are detected before a write to the cache and are stored as a one bit tag per byte (called *Zero Indicator Bit* or *ZIB*). ZIBs require an area overhead of around 9% [221].

Non-zero bytes are stored in their entirety with their ZIB unset and are accessed in the usual manner. A zero byte, however, is only accessed by reading its ZIB. No other bits are read because the ZIB gates their byte wordline signal. Since the storage cells are prohibited from discharging their bit-lines, a compensating circuit must be present in the corresponding senseamps to create zeros without actually reading anything. Depending on the outcome of the ZIB senseamp, the rest of the byte's senseamps are either allowed to detect values on the bit-lines (the byte is nonzero) or are forced to produce zeros (because the whole byte is zero).

Detailed SPICE simulations of this scheme for a 16KB data cache (in a TSMC 0.25 μm CMOS process) show an overall read latency overhead of 2 FO4 (Fan-Out 4) gates and a 26% reduction in dynamic power (10% for instruction caches) for the SPECint95 and Mediabench benchmark suites [221].

4.4.2   Value Compression and the Frequent Value Cache

Significance compression (suppression of strings-of-zeros or strings-of-ones) *is* a rudimentary form of compression. It is a small step to go from there to a compression scheme that can be applied to general values. However, such a scheme cannot be overly complex; no more energy should be expended in compressing and decompressing values than the energy saved by using them. A simple, yet effective compressor must be selected for this purpose.

One solution can be found in *frequent value locality*, a program property, first shown by Zhang, Yang, and Gupta [248] and independently by Larin [148]. Frequent value locality means that a small number of distinct values often accounts for a large portion of the value stream that is accessed. This set of frequently occurring values changes slowly during the execution of a program [130]. Frequent value locality motivated approaches to increase the effective capacity of the L1 cache by packing more compressed cache lines in the cache [237] and to reduce cache power consumption by accessing fewer bits [235, 234]. Both approaches achieve their goal via *frequent value compression*.

In frequent value compression, a *dictionary* is loaded with the frequent values of a program. Every occurrence of a frequent value in the cache is then replaced by an index to the actual value in the dictionary. By virtue of its small size the dictionary can be accessed quickly and provide the actual value with little delay. Typical frequent values include 0, 1, −1, and program-specific values. For example, the SPEC2000 benchmark *perl* very often uses the value 0x78787878 (in the Simplescalar simulator infrastructure). Frequent values are preloaded in the dictionary by profiling the program beforehand. Alternatively, the dictionary enters a short learning period in the beginning of each program until it fills up with values. Dynamic directories that are updated during run-time have been also proposed [130]. The set of useful frequent values is small: just eight frequent values account for 48% of the memory accesses (on average) in several of the SPEC2000 benchmarks.

Frequent value compression is important because it is simple. In terms of power, it is one of the most efficient compression mechanisms. In terms of compression ability it goes well

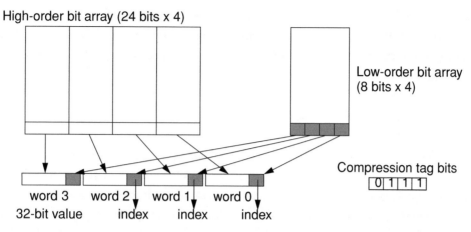

FIGURE 4.9: Frequent value cache (FVC). The low-order bit array provides dictionary indices for the compressed words or the low-order 8 bits of uncompressed words. In the latter case, a second access is required to retrieve the 24 high-order bits. Adapted from [235].

beyond significance compression, but to go any further, complexity and power consumption increase significantly.

Frequent value cache: The frequent value cache (FVC), built around this compression scheme, specifically targets power consumption [235, 234]. In FVC, a cache line can contain both compressed and uncompresed words. Their status is determined by additional bits in the line's tag. A compressed word is simply an index to the frequent value dictionary. The index occupies the low-order bits of the original value leaving the rest of the word empty. Assume that the index occupies the eight low-order bits, allowing for a 256-entry frequent-value dictionary.

The key challenge is how to structure the cache to reduce the energy cost when a compressed word is accessed. FVC does this by splitting cache lines into two different data arrays: the first array (shown on the right in Figure 4.9) holds only the 8 low-order bits of each word in the cache line, while the second array (shown on the left) holds the remaining 24 high-order bits of each of the words. Initially, only the first array is accessed. Thus, only the indices of the compressed words or the eight low-order bits of uncompressed words are obtained. If the requested word is compressed (indicated by the corresponding tag bit) minimal energy was spent to access exactly what was needed—the dictionary index. The dictionary is accessed next to obtain the actual value, but this is not nearly as expensive as accessing the cache. If the accessed word is uncompressed, only its 8 low-order bits are accessed. The rest 24 high-order bits are still needed. For this, the second array is accessed in the subsequent cycle.

Both the dictionary access and the access to the second array increase cache latency, but the dictionary access much less so. Yang and Gupta report an overall increase of about 3% in the execution time of SPEC95 benchmarks but at the same time a 29% reduction of energy for a 64KB L1 cache [235]. The energy reduction comes, of course, from efficiently accessing compressed values. The cost of accessing the smaller low-order bit array plus the cost of accessing the dictionary is less than the cost of accessing the full cache line in a standard cache organization.

4.4.3   Packing Compressed Cache Lines: Compression Cache and Significance-Compression Cache

DZC and FVC save power solely by accessing fewer bits for compressed items. In both techniques, the space freed by compression simply remains empty in the cache. This empty space can be exploited by squeezing more than one compressed lines in a cache frame [237, 141]. This is equivalent to narrow-width operand packing in the datapath and, likewise, is intended to increase cache *utilization*. Besides the power savings of accessing compressed values (reading fewer bits), increasing cache utilization yields indirect power savings since the lower part of the memory hierarchy is accessed less frequently.

☞   *packing techniques*: There are two techniques to pack compressed lines in cache frames: *variable packing* and *fixed packing*.

- *Variable packing* aims to maximize the utilization of the available cache space by compressing and packing a *variable* number of cache lines into a cache frame. Cache line boundaries within a frame are not fixed since cache lines can be compressed to different sizes. This necessitates an indirection for locating the compressed lines [6, 90]. Variable packing is expensive both in terms of power and latency due to the complex compression and placement of lines in cache frames. It is therefore best suited for L2 implementations [6]. The possible power benefit of variable packing in the L2 comes from increasing the hit ratio, not from accessing the compressed lines. Variable packing is not a veritable low-power technique since it is quite expensive on its own. We will not expand further, but for a good discussion of variable packing, see the work by Alameldeen and Wood [6] and the work of Halnor and Reinhardt [90].

- In *fixed packing*, a preset number of cache lines are packed in a cache frame if they can be compressed to fit in pre-allocated spaces. Fixed storage requirements come at the price of reduced opportunities for compression—some cache lines simply do not compress enough to fit. But in contrast to variable packing, an implementation based on fixed packing [235, 234, 141, 130] achieves power benefits not only from fewer accesses to the next cache level, but also from power-efficient accesses to compressed data.

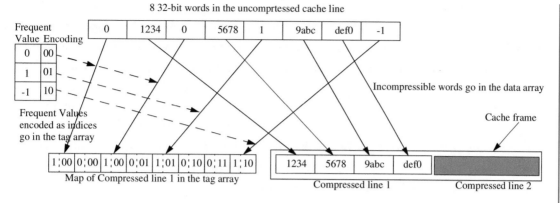

FIGURE 4.10: Frequent value compression for fixed packing. Adapted from [237].

Compression cache: The compression cache (CC) is a fixed packing technique (see "Sidebar: Packing Techniques") proposed by Yang, Zhang and Gupta [237]. Like FVC it uses frequent value compression.[4] Unlike FVC which does not attempt to pack cache lines into frames, in CC each cache frame holds either an uncompressed line or two compressed lines. Two tags per cache frame are provided to handle the latter case. Two lines in the space of one implies that each would have to compress at least to half its size—a difficult feat in most cases. Instead, a little extra space *in the tag array* is provided to facilitate compression.

Here is how it works. Compression is carried out by separating a line's frequent-value words from its incompressible words. The latter are concatenated and stored in the data array, occupying up to one-half of the cache frame (Figure 4.10). Obviously, for this to work, no more that half of the line's words can remain uncompressed but, ultimately, this limits the opportunities for compression. Frequent values are moved to the tag array along with the line's tag. There, they are replaced by their corresponding dictionary indices and stored in a "map" of the uncompressed cache line. The map provides the means to reconstruct the positions of all the words in the compressed line.

Despite CC's performance orientation, it still provides power savings by reducing activity in lower levels of the memory hierarchy. Yang, Zhang, and Gupta show that CC provides energy savings over a direct mapped cache, ranging from 1% to 26% depending on configuration [237].

---

[4]FVC was actually proposed later by two of the same authors.

Significance-compression cache: Following in the footsteps of Yang et al. work, Kim et al. [141] propose a scheme where lines are compressed using sign compression. Again, two compressed lines can fit in the space of one, which means that the compression ratio must be at least 50%. For this to happen, all 32-bit words in a line must be sign-compressed to 16 bits. Although this is possible, it is not very common. To increase the chances of fitting two lines into one, extra space is provided in the data array for words that cannot be compressed by significance compression. According to Kim et al [141], the inclusion of the extra space to accommodate just one incompressible word per line makes a significant difference in the opportunities for compression. Tolerating 25% incompressible data in cache lines yields energy savings on a par with those in CC (about 23%) and for the same reasons.

4.4.4   Instruction Compression

Hines, Green, Tyson, and Whalley proposed the equivalent of the frequent value cache for instructions [100]. They observe that, similarly to data, instructions too exhibit locality: a small number of static instructions appear quite frequently in the dynamic stream. An impressive 80% of the dynamic stream in the MiBench suite can be easily captured with 64 different static instructions. These can be stored in a dictionary—called Instruction Register File (IRF)—similar to the frequent value dictionary. All instances of the frequent instructions in the code can then be replaced by their index to the dictionary. This results in code compression, improved fetch bandwidth, and energy savings. An innovation in their scheme is that the instructions in the dictionary are parametrizable in their immediate field, so a few variations of the instruction appearing in the static code can be covered by a single IRF entry. Hines et al. report that the benefit in performance is not substantial, but the reduction in I-Fetch energy can reach 37% for the MiBench [100] benchmark suite.

4.5   IDLE-CAPACITY SWITCHING ACTIVITY

Idle-capacity switching activity is essentially wasted activity related to out-of-order execution. Similarly to the idle-width activity at the operation level, it appears because processor resources (instruction queues, load/store queues, reservation stations, reorder buffers, caches, etc.) are *over provisioned* to support a high-instruction throughput for target workloads.

Not every program, however, achieves the maximum instruction throughput and utilizes these resources to full extent. This results in excessive power consumption for the achievable performance. By re-balancing—most often *resizing*—processor resources to fit program needs, power consumption can be trimmed to be commensurable to the attained performance. This is the overarching goal of the techniques presented in this and the following three sections.

TABLE 4.4: Estimated Energy Growth Parameters (Adapted From [252])	
Register Rename Logic	$\gamma = 1.1$
Instruction Issue Window	$\gamma = 1.9$
Memory Disambiguation Unit	$\gamma = 1.5$
Multiported Register File	$\gamma = 1.8$
Data Bypass Mechanisms	$\gamma = 1.6$
Functional Units[a]	$\gamma = 0.1$
[a]Power consumed by a functional unit is largely independent of the issue width.	

4.5.1 Power-inefficiency of Out-of-order Processors

What we call idle-capacity switching activity is intuitive to anyone who has studied the workings of an out-of-order processor: large structures, often using content addressable memory, are provided for peak performance, but typically underutilized. To further increase performance, these structures must be enlarged, but the resulting increase in performance pales in comparison to the increase in power and area. It was not until 2000, when Zyuban and Kogge put this intuition into a formalism and showed that out-of-order processors are inherently *power-inefficient* [252].

But what do we mean by "power-inefficient?" Zyuban and Kogge argue that a micro-architecture is *energy-efficient* if its energy consumption increases commensurably with its IPC (instructions per cycle) when the microarchitecture is scaled in its issue width.

To support this formally, Zyuban and Kogge set out and develop energy models for the key structures (register rename table, bypass mechanism, instruction window, load/store window, register file, and functional units) in an out-of-order processor. By analyzing energy consumption as the issue width of the machine is increased, they derive an "energy growth" factor for each structure. This factor shows how the energy-per-instruction *grows* as a function of the issue width. Table 4.4 shows their estimates for the energy growth factor [252].

The *energy-per-instruction* growth for a particular structure can then be described as

$$E_i \sim (\text{IW})^\gamma,$$

where IW is the issue width and γ is the structure's energy growth factor. For instance, the instruction window has the worst γ and its energy consumption grows almost quadratically with respect to the issue width.

Given the definition for the energy growth γ, Zyuban and Kogge show that an out-of-order processor cannot be power efficient given the observed energy growth factors and the achievable IPC at various issue widths. They do this by expressing EDP as a function of issue width scaled by the energy growth.

Here is how it is done. We can express EDP (E × D) as a function of IPC which turns out to be energy-per-cycle divided by IPC^2.

$$E \times D = \frac{\text{energy}}{\text{instruction}} \times \frac{\text{cycles}}{\text{instruction}} = \frac{\text{energy/cycle}}{IPC^2}$$

Substituting energy-per-cycle with the equivalent energy-per-instruction (E_i) multiplied by IPC we get:

$$E \times D = \frac{\text{energy/cycle}}{IPC^2} = \frac{E_i \times IPC}{IPC^2} = \frac{E_i}{IPC} \frac{(IW)^\gamma}{IPC}.$$

If we now assume that IPC is related to issue width via the equation:

$$IPC \sim (IW)^\alpha,$$

we arrive at the following formula expressing EDP as a function of the issue width:

$$E \times D \sim (IW)^{\gamma - \alpha}.$$

So the fundamental question in power efficiency is whether the increase in performance (governed by α) resulting from an increase in the issue width (IW) outweighs the increase in energy consumption (governed by γ). In the real world, α is significantly less than 1. Therefore, any structure with γ greater than 1—and there are almost no exceptions to this—contributes to power inefficiency. The instruction window turns out to be the worst offender.

4.5.2   Resource Partitioning

Two years prior and unrelated to the thread of research on the power inefficiency of out-of-order processors spearheaded by Zyban and Kogge, Albonesi proposed a technique that would form the basis for many of the solutions addressing program-level over-provisioned activity. Albonesi's technique is the resizing of processor structures and at the time it was proposed in 1998, it was not about power [7]. Albonesi's goal was to devise complexity-adaptive structures that traded latency with capacity. In effect, he proposed resizable structures that become faster as they become smaller (or slower as they become larger).

The starting point for this idea is the observation that wire delay increases relatively to gate delay as we go deeper into nanoscale technologies. This, in turn, means that we could not possibly afford the latency of very long wires. To overcome this problem, repeater methodologies

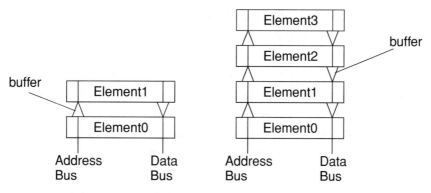

FIGURE 4.11: Complexity-adaptive structures. Adapted from [7].

are used in which a long wire is partitioned into shorter segments by placing buffers at regular intervals along its length. Most of the processor's critical structures are (or will be) in need of such methodologies.

The key observation made by Albonesi is that buffered wires readily lend themselves to *partitioning* without any—significant—additional overhead. In other words, the ability to disable part of a buffered wire (e.g., starting from some point onwards) comes almost for free with buffering. The only catch is that tri-state buffers must be used in this case. Partitioning also circumvents the problem of increased power consumption due to the repeaters (see "Sidebar: Wire Partitioning") by using only as much wire needed.

Large CAM or SRAM structures need long wires for their bit-lines and wordlines. Implementing them with buffered wires endows such structures with the ability to partition and deactivate some part of them—again, virtually for free. Figure 4.11 shows the main idea in such structures.

Albonesi's target was to devise a complexity adaptive architecture where various structures can be resized to provide a size versus speed trade-off. This trade-off is adjusted to best suit application needs. If, for example, an application would benefit from larger structures, even at the expense of a slower clock, a complexity-adaptive architecture would adjust to provide this trade-off. Conversely, applications that have no need for large structures but their performance is tied to clock speed, benefit by scaling down resizable structures to achieve faster clocks. In practice, varying the clock fast enough can be challenging. Moreover, slowing down everything just to increase the size of a single structure can quickly bite into any potential benefits. For these reasons, rather than slowing down the clock, the latency (in cycles) of the resized structures could be increased instead.

Albonesi gives convincing examples by adapting two separate structures, the cache hierarchy and the instruction queue, to the needs of various applications. He did this, however, by examining all possible configurations for these structures and for the application as a whole—although he also mentions finer-grain adaptability at the end. In other words, he did not specify, in this first paper, any mechanisms or policies needed to dynamically adapt the hardware. This is one of the main contributions of the work that followed suit.

☞   *Wire Partitioning*: Because the delay of a wire is proportional to its $R \times C$ product, and both R and C are proportional to the wire length, it follows that the wire delay is proportional to the square of the wire length. Long wires can become unacceptably slow. To solve this problem, a long wire is broken into smaller segments by placing repeaters (inverters or buffers) at regular intervals along its length. The repeaters are driving each of the smaller segments. Breaking a wire into k segments improves wire delay by a factor of k^2 but the repeaters themselves introduce additional latency. However, there is a downside.

The total combined power consumption for driving the segmented wire can be significantly higher than that of the original wire. In fact, the total energy needed for the optimally-sized repeaters increases exponentially with k [39]. A segmentation factor *smaller* than the optimal-delay factor is typically chosen so that the increase of power is kept at bay.

To improve on power, the solution is to actually disable a part of the wire that is not needed. This partitioning technique replaces the inverters or buffers with tristate devices that have the ability to *electrically disconnect* a segment of the wire (see Figure 4.12).

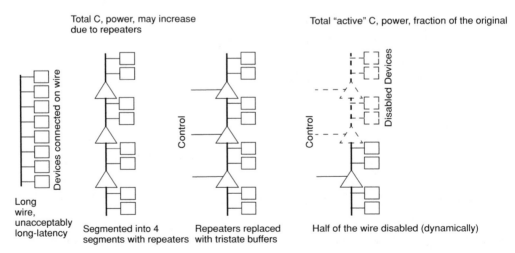

FIGURE 4.12: Wire partitioning. Breaking a wire into multiple segments by inserting repeaters along its length dramatically improves its delay (which is proportional to the square of the wire length) but may end up increasing the total power required due to the additional overhead of the repeaters. Architecturally, this organization can be exploited for low-power by replacing the repeaters with tri-state buffers and dynamically disabling part of the wire (along with everything connected on that part) when this is advantageous.

4.6   IDLE-CAPACITY SWITCHING ACTIVITY: INSTRUCTION QUEUE

Following Albonesi's work on complexity adaptive structures, a number of researchers proposed to dynamically adapt the instruction queue to the needs of programs, but now having power consumption—rather than performance— in their sights. Table 4.5 gives an overview of the proposals we present here.

4.6.1   Physical Resizing

Buyuktosunoglu and Albonesi teamed up with Schuster, Brooks, Bose, and Cook from IBM and published one of the first papers addressing over-provisioned activity in the instruction queue.

TABLE 4.5: Techniques to Adapt the Issue Queue to Program Behavior

Technique	Partitioning	Resizing	Feedback Control
Buyuktosunoglu et al. [42]	Physical partitioning into segments. Partitioning of the IQ is achieved with buffered wires	*Physical resizing*: enabling or disabling consecutive IQ segments	Readiness Feedback: Readiness measured as the number of "active" segments (having the majority of their entries *ready-to-issue*) IPC feedback for limiting performance degradation
Ponomarev, Kucuk, and Ghose [182]	Physical into segments	Physical resizing: enabling or disabling IQ segments	Occupancy Feedback: Instruction Queue Occupancy measured as the fraction of *valid* entries
Folegnani and González [80]	Logical into segments	*Logical* resizing without partitioning: part of the empty space is excluded from allocation to new entries	ILP Contribution Feedback: usefulness of (logical) segments in issuing instructions

The paper discusses a detailed circuit-level implementation of a resizable (adaptive) issue queue (IQ)—conceptually similar to Albonesi's complexity-adaptive design. More interestingly, the paper proposes a high-level mechanism for adaptation [42]. The design of the IQ is a mixed CAM/SRAM design where each entry has both CAM and SRAM fields. The SRAM fields hold instruction information (such as opcode, destination register, status, etc.) and the CAM fields constitute the wakeup logic for the particular entry, holding the input operand tags. Results coming from functional units match the operand tags in the CAM fields and select the SRAM part of the entry for further action. When an instruction matches both its operands it becomes "ready" to issue and awaits to be picked by the scheduler.

The IQ is divided in large "chunks" with transmission gates placed at regular intervals on its CAM and SRAM bit-lines. The tag match in the CAM fields is enabled by dedicated *taglines*. Partitioning of the IQ in chunks is controlled by enabling or disabling the transmission gates in the bit-lines and the corresponding taglines. Figure 4.13 shows the design.

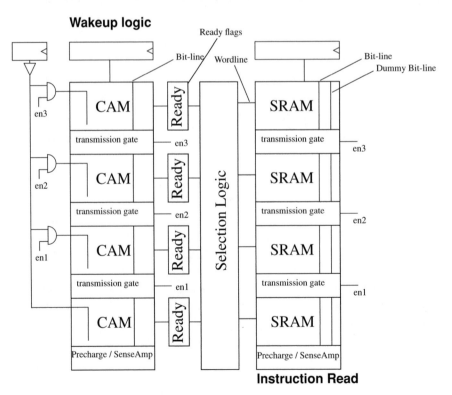

FIGURE 4.13: The design of an adaptive Instruction Queue. Enable signals (en$_X$) enable the CAM and SRAM parts. Adapted from [42].

```
if (present_IPC) < D factor * last_IPC
  increase size;
else if (number of active segments < threshold_1)
  decrease size;
else if (number of active segments < threshold_2)
  retain current size;
else increase size;
```

FIGURE 4.14: Algorithm to resize an Issue Queue. Adapted from [42].

Not surprisingly, the power of the CAM part is linear to the number of entries enabled. If a 32-entry IQ is divided in to four 8-entry chunks, disabling three out of four chunks yields power savings of 75%. The CAM part consumes ten times more energy than the SRAM part, thus the reduction of the CAM energy provides the bulk of the benefit.

4.6.2   Readiness Feedback Control

Perhaps, more interesting than the design of the partitionable IQ, is the algorithm to control its size. The idea is to adjust the IQ size based on the "activity" of its entries. Although the authors do not discuss it in detail, the high-level scheme they propose bases its decisions on the average number of *active* IQ chunks within a time window.

An IQ chunk is regarded as active if at least half of its entries have their ready flag set— i.e., an active chunk has a significant percentage of its entries ready to issue. On every cycle, the number of active chunks is accumulated in a register. At the end of a timing window this register is compared to two empirically chosen thresholds and a decision is taken on whether to disable chunks, lave as is, or enable more.[5]

This decision-making scheme is wrapped with a safety mechanism, that reverses the last (downsizing) decision if it had a negative effect on IPC. The threshold that triggers the safety mechanism is given as a degradation factor D on IPC: if the new IPC is D times the old IPC (where D is less than 1) then the last sizing decision is reversed. The full decision scheme in pseudocode is shown in Figure 4.14.

Using this scheme in a simulated 4-issue processor with a 32-entry issue queue, and on some of the integer SPEC2000 benchmarks, the power savings for the IQ are 35% (on average) with an IPC degradation of just over 4% [42].

4.6.3   Occupancy Feedback Control

Ponomarev, Kucuk, and Ghose examine the more general problem of reducing power for the three main structures that collectively comprise the instruction scheduling mechanism:

---

[5]The authors imply that the thresholds change according to the number of enabled chunks but no further detail is given [42].

the Instruction Queue (IQ), the load/store queue (LSQ), and the reorder buffer (ROB) [182]. As in Albonesi's work [7] (and later in Buyuktosunoglu et al. [42]) these structures are assumed to be physically partitioned into chunks that can be dynamically enabled and disabled.

The major contribution compared to the previous work is the feedback control technique they propose for dynamically resizing these structures. Ponomarev et al. argue that feedback control based on performance metrics such as IPC, or on the number of ready entries, does not reflect the true needs of a program. All these metrics can be misleading since they are not solely affected by the instruction queue size but actually depend on many other factors: cache miss rates, branch misprediction rates, amount of instruction-level parallelism, etc.

Instead, they propose that the *occupancy* of a structure (the percentage of valid entries) is the appropriate feedback control metric for resizing. For the three structures that compose the instruction scheduling mechanisms (IQ, LSQ, and ROB), they show that their occupancies are positively correlated, meaning that all three should be resized in concert. However, because the *ratios* of the occupancies among these structures vary over time, it is imprecise to resize all three based on the occupancy of a single one (e.g., the IQ). Rather, it is better if each structure is resized based on measurements of its own occupancy.

The feedback control scheme periodically samples the occupancy and makes decisions at the end of an update period (which is a multiple of the sampling period). A structure is downsized by disabling as many partitions as can fit in its empty space (which is determined by its occupancy measurements). If, however, a structure fills up and the allocation of new entries is blocked, then it is upsized by powering up more partitions. Upsizing is triggered when the number of cycles that new entries remain blocked exceeds a certain threshold. Upsizing is aggressive to limit performance degradation. In simulations for a 4-issue processor, this method yields power savings for the three structures in excess of 50% with a performance loss of less than 5%.

4.6.4   Logical Resizing Without Partitioning

Folegnani and González take a different path to adapt IQ power[6] to program needs [80]. Instead of physically segmenting the instruction queue and then disabling large chunks at a time, they do the following: first, they disable individual IQ entries; second, they limit the instruction queue's size not physically but *logically* by limiting the part that can be allocated to new entries.

---

[6]According to their estimates, instruction queue power accounts for 25% of the total power consumption of a typical microarchitecture.

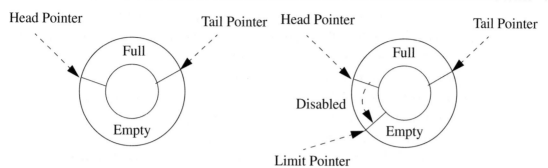

FIGURE 4.15: Circular IQ. Adapted from [80].

The instruction queue in their proposal is organized as a circular FIFO buffer. Head and tail pointers point to the head entry (the oldest instruction) and the tail entry (the newest instruction). The space between the tail and the head entries is the *full* part since it contains valid instructions, either ready to issue or waiting for their operands (Figure 4.15). The space between the head and the tail entries is the *empty* part. Similarly to the IQ discussed previously, CAM fields in each entry match results returning from the functional units. When an instruction matches both its operands, it becomes ready to issue.

☞ *Instruction Queue Collapsing*: Upon issue of an instruction to the execution units, the corresponding instruction queue entry is freed. This creates holes in the *full* part of the IQ (see Figure 4.15). In some designs, such holes are filled by moving up all valid entries. This is called *collapsing* and it is done because it can simplify the selection (scheduling) of ready instructions. One example of this technique is the instruction queue of the Alpha [134]. However, collapsing consumes power because of all the data movement it entails. Folegnani and González do not use it since holes in the full part are also included in their power-saving schemes.

The key observation Folegnani and González make about such an instruction queue is that empty entries need *not* participate in the tag match at all. Furthermore, ready operands also do not need to participate in the tag match. It is fairly straightforward to disable an entry's CAM tag comparison by gating the tagline precharge transistor with the entry's ready flag or valid flag. This immediately reduces the comparison activity, making it proportional to the number of *valid waiting* entries in the IQ. According to their statistics for an 128-entry IQ and for representative SPEC2000 benchmarks, on average, there are only 58 entries in full area of the IQ and 26 of those are already empty. This means that about 89% of the wake-up energy (CAM matching) can be saved. Similarly to the estimates of Buyuktosunoglu et al., Folegnani and González also attribute the bulk (63%) of the IQ power to the associative matching.

Despite the magnitude of these savings—the low-hanging fruit in this case—Folegnani and González go one step further. They resize the IQ to fit program needs. The interesting difference from the earlier proposals is the IQ is resized *logically*—not physically—by partitioning and selectively disabling parts of it. Besides the head and tail pointers, they introduce a new pointer, called the *limit* pointer which always moves at a fixed offset from the head pointer. This pointer limits the logical size of the instruction queue by excluding the entries between the head pointer and itself from being allocated. Figure 4.15 shows the new "disabled" area defined by that pointer. What this does is to add a known number (the offset from the head pointer) of guaranteed empty entries that will not participate in tag matching. The question now is how to maximize the disabled area without negatively impacting performance.

ILP-contribution feedback control: This is done using a heuristic with empirically chosen parameters. The IQ is logically divided in 16 partitions. The idea for the heuristic is to measure the contribution to performance from the youngest partition of the IQ which is the partition allocated most recently at the tail pointer. The contribution of a partition is measured in terms of issued instructions from this partition within a time window. If that contribution is below some empirically chosen threshold, then the effective size of the IQ is reduced by expanding the disabled area. Periodically the effective IQ size is increased (by contracting the disabled area). This simple scheme increases the energy savings to about 91% with a modest 1.7% IPC loss.

4.6.5 Other Power Optimizations for the Instruction Queue

About the same time as with the Buyuktosunoglu et al. and the Folegnani and González papers, a slew of techniques were proposed to reduce power in instruction queues. Some of them are at the circuit level but motivated by architectural characteristics, such as the techniques proposed by Kucuk, Ghose, Ponomarev, and Kogge [145]. They propose three techniques to reduce IQ dynamic power: (i) efficient comparators in the CAM part, (ii) significance compression in the SRAM part (which is another example of the technique described in Section 4.3, "Idle-Width Switching Activity: Core"), and (iii) Bit-line segmentation (which is explained in "Sidebar: Bit-line Segmentation"). Their first proposal goes beyond disabling the tag match for empty entries and ready operands. It attempts to minimize tag match energy for all those entries that do participate in the tag comparison but not match. The technique exploits the prevailing per-bit behavior of typical programs. Because of the localization of dependencies in a program, a mismatch is much more likely to occur in the least significant bits of an operand tag. This means that just checking the lower four out of the eight tag bits can reveal a mismatch of 90% of the time saving, in this case, half of the power of a full comparison.

The structures that comprise instruction scheduling, especially the instruction queue, have been prime targets for power optimizations. And for a good reason: not only instruction scheduling consumes a significant part of a processor's total power, but also does not scale well to larger sizes. Starting from the techniques discussed here, many others followed in the literature offering further improvements and optimizations.

4.6.6   Related Work on Instruction Windows

The importance of the instruction window for performance, its high complexity, and its criticality (in terms of latency), make it a prime target for optimizations. Here, we give a short overview of some of the most relevant work. Even though this work is mostly focused on performance or complexity and does not specifically addresses power consumption, it can have significant ramifications on power.

Palacharla, Jouppi, and Smith, in the context of their work on complexity-effective architectures [177], first proposed to lower the complexity of a CAM-based monolithic instruction queue by replacing it with a number of FIFO queues. Instead of searching the whole IQ for ready instructions, the search is limited to the heads of the FIFOs. Subsequently, Canal and González [45, 46] and independently Michaud and Seznec [166] propose dataflow scheduling arrays, augmented with fully associative buffers to accommodate unpredictable-latency instructions.

Following these initial proposals on reducing the complexity of an ordinary sized instruction queue, the converse idea came into focus: instead of reducing complexity, apply these techniques to actually *enlarge* the instruction queue. Two groups, Raasch, Binkert, and Reinhardt and, independently, Lebeck, Koppanalil, Patwardhan, and Rotenberg, proposed to enlarge IQ size using dependence chains [186, 149]. Dependent chains are groups of dependent instructions usually headed by an instruction of unpredictable latency. A head instruction can itself be a dependent instruction in another dependence chain.

Raasch et al. proposed a segmented IQ design where instructions are promoted from segment to segment until they reach the "issue-buffer" segment. From there (and only there) they can issue to the functional units. Instructions are placed in the various IQ segments according to their *expected delay* in becoming ready. For example, ready instructions and their immediate dependent instructions are placed in the first (issue-buffer) segment; instructions that are two or three cycles away from becoming ready are placed in second segment (after the issue-buffer); and so on. In each cycle instructions are promoted from segment to segment as they advance toward becoming ready.

The important innovation in their proposal is that the expected delay of instructions is not measured in absolute time but *relatively* to the head of the dependence chain. Thus, as the head is promoted from segment to segment, it pulls behind it like the wagons of a train its chain of dependent instructions. A large number of small segments allows for a very large instruction queue while at the same time for *faster* clock speeds. However, compared to a monolithic design of the same size (and assuming equal frequency), the dependence-chain IQ delivers somewhat less performance. On the other hand, because the part of the IQ involved in issuing instructions is quite small, power consumption can be substantially less than a monolithic design.

Lebeck et al. also propose a technique based on dependence chains [149]. In their scheme, the instructions depending on a long-latency operation (e.g., a cache miss) are moved out of the issue queue into a much larger "waiting instruction buffer" (WIB). A 2K-entry WIB with a 32-entry IQ yields noteworthy speedups over a conventional 32-entry IQ. Although a power analysis was not included in their work, it is likely that the power benefits for such a design would also be noticeable compared to a (large) monolithic IQ at the same performance level.

Similarly, hierarchical scheduling windows (HSW) [34] increase the instruction window size using a fast (but small) and a slow (but large) scheduling window. Latency-sensitive critical instructions are moved to the fast scheduling window while latency-tolerant instructions remain in the larger slower window. The difference from the previous work is that both scheduling windows issue instructions, each to a separate cluster.

Other proposals are focused on reducing the cost—and by extent, complexity and power— of checkpointing, which is a serious impediment to large instruction windows [5, 82]. Finally, there are also proposals that employ segmentation and resizing as well as more targeted optimizations to reduce the design complexity of load/store queues [179]. All of these techniques are *power-efficient* in the sense that they attempt to increase performance but with a very prudent and frugal use of resources.

4.7 IDLE-CAPACITY SWITCHING ACTIVITY: CORE

Besides the instruction window, significant opportunities still abound in a dynamically scheduled out-of-order processor for further power optimizations. A related dimension to the instruction window size is the issue width—the number of instructions that can go through the processor in parallel. Although we talk about issue *width*, we consider such techniques under the *Idle-Capacity* optimizations. The reason is that, in contrast to idle-width optimizations discussed in Sections 4.3 and 4.4, adapting the issue width has nothing to do with the bit-width of individual operations (arithmetic, logic, or memory operations) but rather with the behavior of the program at a larger scale. This is consistent with, and in fact very similar to, the other idle-capacity optimizations presented in Sections 4.5 and 4.8.

Depending on the instruction window size different programs exhibit different maxima for parallel instruction issue. Adapting the processor to this dimension was proposed by Bahar and Manne [19]. They propose to dynamically change the width of an 8-issue processor to

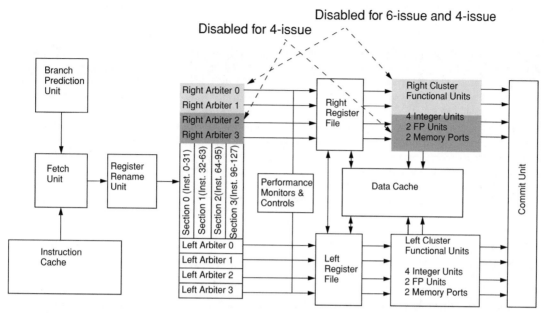

FIGURE 4.16: Adjusting the width of an 8-issue machine to 6- or 4-issue. Adapted from [19].

6-issue or 4-issue when the application cannot take advantage of the additional width. They model their target processor after an 8-issue Alpha 21264 [134], comprising two 4-issue clusters (Figure 4.16). To switch the processor to 6-issue, one-half of one of the clusters is disabled. To switch to the 4-issue, one whole cluster is disabled.

To disable half or a whole cluster, the appropriate functional units are clock gated. In addition to disabling functional units, part of the instruction queue hardware is also disabled, thus realizing additional power benefits. This is different from IQ resizing (described previously), in that the instruction window *size* does not change. What is disabled in the IQ are the scheduling units that pick ready instructions for execution (called *arbiters* in Alpha parlance). All in all, this technique can save up to 20% (10%) power from the execution units, 35% (17%) from the instruction queue, and 12% (6%) in total, in the 4-issue (6-issue) low-power mode.

Again, one of the main contributions of this work is the technique to adjust the issue width to the needs of the program. Decisions are made at the end of a sampling window assuming that the behavior of the program in the last window is a good indicator for the next.[7] A set of rules decides whether to enable or disable the two low-power modes, the 4-issue and the 6-issue mode. Enabling a mode sets the width to 4 or 6, but disabling it jumps back to normal operation, which is the full 8-issue mode. The rules compare the issue IPC (for both

---

[7]This assumption is put to test by Bahar and Manne and is shown to hold up quite nicely [19].

TABLE 4.6: Rules for Switching to Low-power Modes. Adapted from [19].

Trigger	Threshold Values
Enable Condition 4-issue	$(I_{IPC} < 3.0)$ AND $(FP_{IPC} < 1.4)$ with hysteresis of two windows
Disable Condition 4-issue	$(I_{IPC} > 3.2)$ OR $(FP_{IPC} > 1.6)$
Enable Condition 6-issue	$(I_{IPC} < 4.5)$ AND $(FP_{IPC} < 1.4)$
Disable Condition 6-issue	$(I_{IPC} > 5)$ OR $(FP_{IPC} > 1.6)$

integer and floating IPC) to a set of empirically derived thresholds. There is also hysteresis for the decision to go to the lowest power mode (the 4-issue mode) requiring the enabling conditions to hold for two consecutive sampling windows. The actual rules used are listed in Table 4.6, where I_{IPC} and FP_{IPC} are the issue IPC for the integer and the floating point instructions, respectively.

These rules for adjusting the width of the machine result in minimal performance loss (on the order of 1–2%) for both single applications and multiprogramming workloads while realizing at times satisfactory per-component power savings (on the order of 10–15%). However, in the big picture, the power savings for the whole processor are not as dramatic. Bahar and Manne point out that a single technique alone cannot solve the power consumption problem. Rather, by taking a holistic approach and applying many techniques in concert, power consumption can be brought down significantly.

4.8   IDLE-CAPACITY SWITCHING ACTIVITY: CACHES

Similarly to instruction queues, caches can also be *sized* to save power. Although it is always better to have more cache, in many cases using much less can save considerable power while giving up very little in performance.

Considerable work is devoted to the general problem of reducing dynamic power in caches. Here, we are only concerned with techniques that *resize* the cache to fit program needs. Techniques that optimize cache access but do not really change the amount of cache available to the programs are discussed extensively in Section 4.9. We concentrate on four proposals that have inspired further work: a cache resizing proposal that trades memory between two cache levels [7, 21], the *selective cache ways* proposal [8] which resizes a cache by changing its associativity, the *accounting cache* proposal [68] which is a combination of the two previous proposals, and finally a CAM-tag cache resizing technique [244]. Table 4.7 summarizes the

TABLE 4.7: Four Approaches for Adapting the Cache to the Needs of Programs

	Partitioning Granularity	Partitioning Technique	Configuration Method
Variable L1/L2 division Albonesi [7], Balasubramonian et al. [21].	Physical Memory Segment (cache bank)	A single physical cache partitioned into segments hosts a virtual two-level cache. Memory segments are allocated to either a virtual L1 or a virtual L2	Configuration searches are initiated on program phase changes. A configuration search goes through each possible configuration and measures CPI, The "best" configuration is selected and persists for the duration of the phase
Selective Cache Ways, Albonesi [8]	Associative Way (may comprise one or more banks)	Individual associative ways can be turned off adapting the size of the cache to program needs	Left to software. Profiling and performance tools indicate the number of ways for some tolerated level of performance-loss
Accounting Cache, Dropsho et al., Albonesi et al. [68, 9]	Associative Way (may comprise one or more banks)	Similar to Variable L1/L2 division, but with the partitioning granularity of the Selective Cache Ways	Performance and energy for all possible configurations are estimated periodically with MRU information, One-shot configuration: the "best" configuration is selected and persists for the next period
Miss Tag Resizing for CAM-Tag Caches, Zhang and Asanović [244]	Individual Set resizing in its associativity	Bit-line Segmentation ("Sidebar: Bit-line Segmentation")	Performance-based Feedback Control: difference in number of misses from the full-sized cache within a time window

main characteristics for the three approaches. Equally important to the partitioning technique is the method for selecting a cache configuration to achieve power or performance goals.

4.8.1   Trading Memory Between Cache Levels

Cache resizing was also proposed in Albonesi's paper on complexity-adaptive structures along with instruction queue resizing [7]. Both techniques rely on structures partitioned in segments using buffered wires. Regarding caches, the whole memory comprising the cache hierarchy is assumed to be segmented in this manner.

Albonesi's proposal calls for a variable division between the L1 and the L2. This dynamic division is based on assigning memory segments to be either in the L1 or in the L2. Architecturally, the two caches are resized by increasing or decreasing their associativity—not by changing the number of sets. Thus, cache indices remain the same throughout size changes. This is necessary to avoid making resident data inaccessible after a change in indexing. Furthermore, cache exclusion is imposed between the L1 and the L2, guaranteeing that data remain unique regardless of the movable boundary between the two levels. Cache inclusion, on the other hand, can result in the same data appearing twice in the same cache. This is possible if two copies of the same data initially residing in the L1 and the L2, respectively, end up in the same cache after a resizing operation.

The variable boundary between L1 and L2 is intended for performance reasons. Making the L1 smaller allows for a faster clock (the latency of the cache in cycles does not change), while making it larger increases its hit ratio. In this initial work, no attempt is made to dynamically control the configuration of the caches. Instead, all possible configurations are studied, each persisting throughout the execution of a program.

Although this complexity-adaptive scheme yields performance benefits (depending on the program and the configuration) no assessment is provided regarding its impact on power consumption. However, the change in associativity in the L1 and the L2 (magnified by the difference in the number of accesses between the two caches) *can* affect power consumption, despite the fact that total amount of active memory remains constant.

Following the initial proposal for the variable L1/L2 division, Balasubramonian, Albonesi, Buyuktosunoglu, and Dwarkadas take it one step further by proposing a more specific and more detailed cache organization to achieve the same goal [21]. More importantly, they also propose mechanisms to control the configuration of the caches at run-time.

The organization is based on a 2MB physical cache which is partitioned into four distinct 512KB subarrays. Each subarray is further partitioned into four segments with the help of repeaters in the wordlines. Each of these segments acts as an associative way, either allocated to the L1 or to the L2. Figure 4.17 shows the organization of the physical cache.

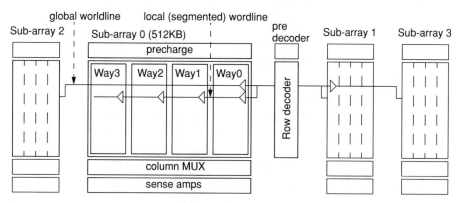

FIGURE 4.17: Physical cache organization. Adapted from [21].

The physical cache hosts a *virtual* two-level hierarchy. Virtual L1 and L2 caches are created within the physical cache by assigning ways to each level. Table 4.8 shows the possible assignments, along with the resulting size, associativity, and access time (in cycles) of the virtual L1. An important difference from Albonesi's first proposal which advocated changing the clock frequency to suit a faster or slower L1 [7], is that the clock frequency remains fixed. What changes is the access latency, in cycles, for both the L1 and the L2. Latency changes in half-cycle increments, assuming that data can be captured using both phases of the clock as in the Alpha 21264.

Similarly to the first proposal [7], the virtual caches are exclusive. On each access, one of the subarrays is enabled by predecoding a *Subarray Select* field in the requested address. Within the enabled subarray, only the L1 section (the shaded ways in Table 4.8) is initially accessed. In case of a miss, the L2 section is then accessed. If there is a hit in the L2, the requested data are moved to the L1 by swapping places with the data already read during the L1 miss. If there is a miss in the L2, data returning form memory is put in the L1 section; any displaced L1 data are moved into the L2 section.

Feedback control: configuration searches: The justification behind this L1/L2 partitioning is that it can adjust to different tolerances for hit and miss latencies. For programs, or better yet program *phases*, that have a very low tolerance in hit latency, a fast L1 can be employed even if it does not yield a very high hit rate. On the other hand, if a program (or program phase) can tolerate somewhat higher hit latency but cannot tolerate a large miss latency, then a larger L1 (albeit somewhat slower) might be the right solution.

The goal is therefore to find a configuration of the virtual caches that yields the right balance between hit latency and miss rate, *per program phase*. Balasubramonian et al. propose a method to achieve this balance but leave open the choice for a software or a hardware implementation. Their method works as follows. Performance statistics (miss rate, IPC, and

TABLE 4.8: Partitioning of the Physical Cache into Virtual L1 and L2 Caches. Shaded Configurations (on the left side) are the Low-power Alternatives for a Specific Size

L1 Size	L1 Assoc	Acc Time	Sub-array 2				Sub-array 0				Sub-array 1				Sub-array 3			
			W3	W2	W1	W0	W3	W2	W1	W0	W0	W1	W2	W3	W0	W1	W2	W3
256K	1 way	2	L2	L2	L2	L2	L2	L2	L2	L1	L1	L2	L2	L2	L2	L2	L2	L2
512K	2 way	2.5	L2	L2	L2	L2	L2	L2	L1	L1	L1	L1	L2	L2	L2	L2	L2	L2
768K	3 way	2.5	L2	L2	L2	L2	L2	L1	L1	L1	L1	L1	L1	L2	L2	L2	L2	L2
1024K	4 way	3	L2	L2	L2	L2	L1	L1	L1	L1	L1	L1	L1	L1	L2	L2	L2	L2
512K	1 way		L1	L2	L2	L2	L1	L2	L2	L1	L1	L2	L2	L1	L1	L2	L2	L1
1024K	2 way	3.5	L2	L2	L1	L1	L1	L2	L2	L1	L1	L2	L2	L1	L1	L2	L2	L1
1536K	3 way	4	L1	L1	L1	L1	L1	L1	L2	L1	L1	L2	L1	L1	L1	L1	L2	L1
2M	4 way	4.5	L1	L1	L1	L1	L1	L1	L1	L1	L1	L1	L1	L1	L1	L1	L1	L1

branch frequency) are gathered in an interval on the order of 100 000 instructions. The statistics produce two pieces of information: first, the CPI (cycles per instruction) in the current window; second, an indication on whether a phase change occurred. A phase change is detected if the statistics in the current window are markedly different from the ones in the previous window. In such a case, any previously selected configuration is discarded and a configuration search starts anew. The sensitivity of the phase detection mechanism is adjusted dynamically so as to not get stuck in a single configuration nor constantly initiate new configuration searches for no good reason.

The search goes through the possible configurations, using each one for a whole time window. The search starts with the 256KB 1-way L1 and progresses through the configurations in Table 4.9, in order. The configuration search also stops if the miss rate drops below some threshold (set to 1% in the paper). Each configuration that is tried out in a search yields a CPI, which is stored in a table. When the search completes (either by running out of configurations or by bringing the miss rate below the threshold) the configuration with the *lowest* CPI is picked. This configuration is called "stable" and persists for the duration of the program phase.

Balasubramonian et al. report on the performance and power consumption of their proposal using a subset of the SPEC95, SPEC2000, and Olden benchmarks [21]. A dynamic L1/L2 division yields no results on programs that have very small miss rates in the L1. But for programs exhibiting a significant miss rate with a conventional 64KB 2-way L1, a dynamic L1/L2 division can improve the CPI by 15% on average (and for some programs up to 50%). This performance improvement, however, comes at a cost: a significantly higher (over 2x) energy per instruction (EPI) for some programs. The reason is that the L1, in the best performing configurations, is highly associative. A low-power modification to the search—selecting the lowest associativity for a specific size—improves the situation by trading some performance improvement for a significant reduction in energy. Projecting to 35 nm technologies and a 3-level cache hierarchy, Balasubramonian et al. show a 43% energy reduction compared to a standard cache.

4.8.2   Selective Cache Ways

One of the key notions in Albonesi's initial complexity-adaptive proposal is that caches can be resized by changing their associativity [7]. In parallel with the variable L1/L2 division proposals [7, 21], Albonesi proposes a much simpler technique, specifically for reducing power consumption. This technique, called "*selective cache ways*" abandons the variable L1/L2 division and concentrates on resizing a single cache by changing its associativity [8].

The idea of selective cache ways is rooted on two observations. First, not all the cache is needed all the time by all programs. In many situations, a smaller cache does (almost) as well, consuming far less power. Second, and equally important, resizing the cache can be done

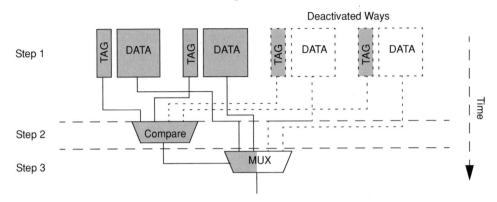

FIGURE 4.18: Selective Cache Way: use smaller associativity. Shaded parts consume power on an access. Deactivation mechanism not shown.

in a way that does not compromise the performance of the full-blown configuration. Unlike *intrusive* approaches that could result in decreased clock speed or increased cache latency (in cycles), the proposal for selective cache ways does not require anything that is not already there.

Large caches are partitioned into multiple subarrays for performance. This is a *design time* partitioning, distinctly different from the dynamic (repeater-based) partitioning mentioned so far. The goal of having smaller sub-arrays in the first place is to break the bit-lines and wordlines into smaller segments to avoid the excessive delays of long wires. The geometry of the sub-arrays is chosen in such a way as to equalize the delay of their bit-line and wordline segments. Given this partitioning, the ways in a set-associative cache typically comprise one or more *independent* sub-arrays. This, in turn, allows complete control over individual cache ways. Unneeded ways can be disabled with little impact on the operation of the rest of the cache.

Disabling a cache way means that its data array simply does not react to cache accesses: its bit-lines are not precharged, its wordlines are not activated, and its sense amplifiers are prevented from firing. The tags of a disabled way, however, remain active. Figure 4.18 shows a block diagram of a 4-way cache where two of its ways have been disabled. The output multiplexor must be configured accordingly to ignore disabled ways.

Disabling a cache way brings up the question of what happens to its data—especially, its modified (dirty) data. A low-power cache architecture cannot exclude a write-back policy, therefore, dirty data are bound to exist in this case. One solution would be to flush the disabled way and write back all its dirty data to memory. This, however, is the expensive solution. The solution adopted in selective cache ways pushes the responsibility (and the complexity) to the cache controller. Data in a disabled way can be accessed by briefly reinstating it into active status. This happens in two situations. First, when a coherence request needs data from a disabled way; second, when there is a hit in a disabled way. Both cases are detected by the cache controller since all tags remain active at all times. In both cases, the relevant data are moved out of the disabled way—temporarily enabled for this purpose—and moved to an enabled way.[8]

Feedback control: Software: Albonesi puts the burden of deciding the appropriate cache configuration on software. First, a threshold for the tolerated performance degradation (called PDT) is established. Any cache configuration that crosses this threshold is not allowed. Performance evaluation tools can then predict how many cache ways an application really needs so as not to lose more performance than what can be tolerated.

However, no attempt is made to evaluate this software-driven strategy. Instead, the evaluation in the paper exhaustively examines all configurations for various cache sizes and PDTs for a subset of the SPEC and reports the results. For energy savings, the combined energy of the selective ways L1 *and* of the L2 must be taken into account. Examining the L1 in isolation might give the wrong picture, since the increase in L2 energy can outpace the savings in the L1. Indeed, in many programs, there is an inflection point in the energy curves. As more and more ways are turned off, after a point, the combined L1/L2 energy consumption begins to rise. The inflection point is specific to each program. Thus, the effectiveness of the technique depends on how much of its cache a program can give up for a given level of performance degradation.

4.8.3   Accounting Cache

Although the idea of selective cache ways has proven to be quite influential, lack of a convenient mechanism to adapt to program needs *at run-time* limits its applicability. Profiling can be an acceptable solution in some cases, for example, in embedded applications. This deficiency of the initial proposal is addressed in the work of Dropsho et al. [68, 9].

Their proposal, termed the *accounting cache*, is a cross between selective ways cache and the proposals for a variable L1/L2 division. It takes its name from the run-time accounting that drives its configuration. The accounting cache is a set-associative organization which—similarly to the selective cache ways—can "disable" a number of its ways. The disabled ways, however, are not inaccessible but simply activated in the case of a miss in the "enabled" ways. Similarly to the proposals for variable L1/L2 division, the disabled ways form a "fake" L2, or rather, a secondary, slower, L1. The enabled ways—the "primary group"—respond first, while the disabled ways—the "secondary group"—respond only after a miss in the primary group.

The access protocol for the accounting cache guarantees an important property for its replacements: that no matter what the division between the primary and the secondary group, replacements follow a global (true) LRU order encompassing all of the ways in a set. This is critical for the run-time accounting proposed in the paper.

---

[8]In the case of a coherence action, the data could be simply discarded if not longer needed.

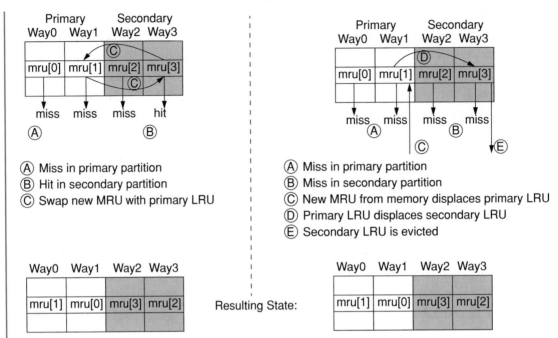

FIGURE 4.19: Replacements in the Accounting Cache. The set MRU order is given by the indices (MRU[0] is the MRU while MRU[3] is the LRU). Adapted from [68].

☞ *global LRU implementation:* Maintaining a global LRU order in this cache is accomplished by placing the most recently accessed blocks in the primary group and moving the displaced blocks in the secondary group (Figure 4.19). For example, a miss in the primary group but a hit in the secondary group results in swapping the primary LRU block with the secondary block that hit (Figure 4.19, left side). A miss in both groups results in a new block brought in from memory directly in the primary group. This displaces the primary LRU block into the secondary group, where it evicts the secondary—global—LRU block (Figure 4.19, right side). This is a viable replacement policy for a *non-inclusive* two-level hierarchy and is used in the original proposals for a variable L1/L2 division.

Feedback control: one-shot configuration and accounting: The replacement policy in the accounting cache guarantees a true LRU ordering in the sets. Based on this ordering, global statistics for the whole cache concerning the number of *hits per LRU position* (or equivalently *MRU position*) can be easily collected using a set of global counters. Thinking in terms of MRU, there are N—where N is the total associativity—global counters $MRU[0]..MRU[N-1]$. $MRU[0]$ is the most recently used and $MRU[N-1]$ is the least recently used. Each time a hit takes place on a line in MRU position i the corresponding counter $MRU[i]$ is incremented.

These statistics are important because hits in various MRU positions correspond to hits in different cache configurations. Hits in the first MRU position correspond to hits in a direct-mapped cache; the combined hits in the first and second MRU position correspond to hits in a two-way set-associative cache; and so on. Thus, hits in *any configuration* of the primary and secondary groups can be derived simply by summing up hits in the appropriate MRU positions. This leads to one-shot configuration by allowing one to *assess* in one go all possible outcomes and select the "best" configuration. In contrast, a configuration search would have to try each and every configuration for an entire interval and then make a decision.

Here's how one-shot configuration is done in more detail. Statistics are gathered in intervals of 100 000 instructions. Since the statistics are independent of the cache configuration in the interval, they can be used to try "what if" scenarios for *any* cache configuration. Assuming that the statistics of an interval are a good indication for the behavior of the next, the most appropriate configuration for the next window can be thus uncovered.

The "what if" scenarios use simple memory access latency and energy cost models. These models calculate the effective memory latency and the energy of a configuration as a function of the hits in its primary and secondary groups. The calculations are performed in a software interrupt handler which also decides on the next configuration.

The policy to decide the next configuration is to go for the lowest energy consumption given a limit in the tolerated performance loss (called *tolerance level*). This sounds similar to the policy used in the selective cache ways, but goes further. It has memory. It keeps an account of what happens in each interval and builds credit or debit for both performance and energy. So, for example, if previous configurations had better performance than the corresponding estimates indicated, the policy becomes more aggressive in trying to reduce energy since it has *performance credit*. On the contrary, if a performance deficit from previous configurations was accumulated, the policy has to make up for it, giving up on energy reduction.

This accounting scheme is a result of the one-shot configuration relying on an *estimate* on what happens in the upcoming interval. This estimate relies, in turn, on the assumption that the measured statistics do not differ noticeably from interval to interval. But in reality they do differ. Accounting normalizes the differences between the estimated and the actual by employing credit or debit in the next configuration decision.

The accounting cache yields very good power results with a rather small impact on performance. As Figure 4.20 shows, for tolerance settings of 1/64, 1/16, and 1/4 (1.5, 6.2, and 0.25 in the graph), energy savings range from 54% to 58% for the instruction L1, 29% to 45% for the data L1, and 25 to 63% for a unified L2 with parallel tag/data access. Overall, for

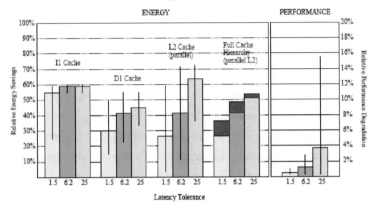

FIGURE 4.20: Accounting cache results. From [68]. Copyright 2002 IEEE.

both cache levels, savings range from 35% to 53% depending on the tolerance settings while performance loss ranges from less than 1% to less than 4%.

4.8.4 CAM-Tag Cache Resizing

Zhang and Asanović point out that while considerable effort is expended on proposals to resize high-performance RAM-tag caches, many real low-power processors actually use the more power-efficient CAM-tag design (see, for example, Section 4.2.4 for a commercial CAM-tag cache). Resizing, however, is equally—if not more—advantageous for highly-associative CAM-tag caches. To fill this gap, Zhang and Asanović proposed the first technique for CAM-tag cache resizing, called *Miss–Tag Resizing (MTR)* [243].

There are distinct advantages in resizing a highly-associative CAM-tag cache. To begin with, resizing at a granularity of an associative way is finer-grain in CAM-tag caches than in RAM-tag caches. The latter have fewer but larger ways (see, for example, the Selective Cache Ways in Section 4.8.2). Moreover, in CAM-tag caches resizing can be done *individually per set*, meaning that an associative way need not be disabled in its entirety across all sets. Rather, in each set, one can disable any associative way without regard to what is happening in other sets. To do this, Zhang and Asanović rely on the bit-line segmentation technique (see "Sidebar: Bit-line Segmentation"). But for bit-lines, there is an important difference between RAM-tag and CAM-tag organizations.

Bit-lines in RAM-tag caches run across the *sets* of the cache. In contrast, bit-lines in CAM-tag caches run across the *ways* of a set. This is shown in Figure 4.21 for a RAM-tag cache of N sets and 4 ways and a CAM-tag cache of N ways and 4 sets. This figure shows that bit-line segmentation in a RAM-tag cache partitions the sets while in a CAM-tag cache partitions the ways. It is also evident, that in the CAM-tag organization each set can be resized

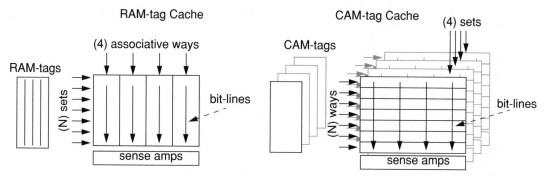

FIGURE 4.21: Bit-line organization in RAM-tag and CAM-tag caches. Bit-lines in RAM-tag caches run across sets. Bit-lines in CAM-tag caches run across ways (not sets).

(in the number of its ways) independently of any other set. This holds regardless on whether a set corresponds to a single bank (as depicted in Figure 4.21) or shares a bank with other sets.

Using bit-line segmentation, a set can be partitioned into a small number of partitions, each encompassing a number of ways. For example, partitioning a 32-way set into 8 partitions, results in 4 ways (cache lines) per partition. In the Zhang and Asanović work [243], resizing is performed in steps of a single way (a single cache line) at a time. Gated-V_{dd}—a technique to reduce leakage—deactivates individual cache lines. However, only when all the cache lines (ways) in a partition are deactivated can the bit-line segment of that partition be taken off the global bit-line and all switching activity in the partition cease. In the example of the 32-way, 8-partition cache, only when all four cache lines of a partition are deactivated is the partition itself deactivated.

The control policy for resizing is a classic *performance-based* feedback loop. What is measured is whether resizing (upsizing or downsizing) leads to worse, better, or the same performance. Depending on the measurement, resizing (up or down) is continued, reversed, or postponed. The metric for performance is the number of misses in a time window of 128K references [243]. One way to use this number would be to compare it to the number of misses measured in the previous time window. But this would entail considerable uncertainty in gauging the effect of a resizing decision on performance. This is because miss rate can vary, not solely as a function of cache size, but as a result of program behavior as well.

Zhang and Asanović propose something more reliable—albeit, at an additional cost. They compare the number of misses of the resized cache to the number of misses that *would arise* in the full-size cache. They do this by keeping a second (full) set of tags, called the *Miss Tags*, whose sole purpose is to count the number of misses in the full-size cache. Miss Tags are only accessed during misses. Since they serve only an accounting role they are not performance-critical

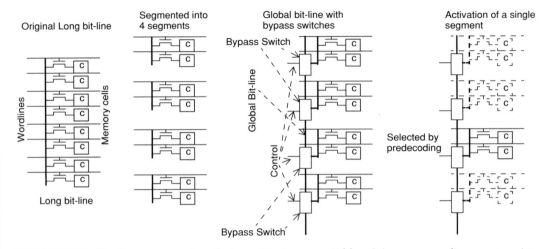

FIGURE 4.22: Bit-line segmentation. Only one bit-line per cell (c) and the corresponding pass transistor are shown. The original bit-line carrying 8 cells is partitioned in 4 segments each carrying 2 cells. The 4 segments attach to a new global bit-line via bypass switches. The bypass switches control dynamically which segment drives the global bit-line. Adapted from [83].

and can be easily made power-efficient. This leaves the area overhead as the main cost for the Miss Tags which Zhang and Asanović estimate to be about 10% for a 32KB cache [243].

The performance difference from resizing the cache is estimated as the difference of the misses of the resized cache and the full cache. The policy is to downsize the cache as much as possible without letting this difference exceed an empirically derived limit. Zhang and Asanović report that MTR yields a significant reduction in the dynamic energy of the order of 28% for the data cache (34% for the instruction cache) compared to a full-size CAM-tag cache.

☞   *bit-line segmentation*: *Bit-line segmentation* applies to long bit-lines in SRAM arrays [125, 83]. Such bit-lines are connected via pass transistors to a large number of memory cells. The capacitive load due to the diffusion capacitance of the pass transistors in addition to the capacitance of the wire itself, significantly adds to the power (e.g., precharge power and sense power) needed to drive such bit-lines. The solution is to break the bit-line in k segments, each carrying a fraction of the cells of the original bit-line (see Figure 4.22). In contrast to wire segmentation, bit-line segments are *not* stringed together with repeaters. Instead, a new global bit-line is introduced to carry the result from each of the bit-line segments.

Each segment of the original bit-line attaches to the global bit-line via a bypass switch (Figure 4.22). These switches dynamically control which bit-line segment drives the global bit-line. Part of the address is predecoded to allow only the relevant bit-line segment on the global bit-line. The power benefit in this case comes from activating only a small segment of the original bit-line. On the other hand, a new wire (the global bit-line) and a number of bypass switches are introduced in the design. The capacitive load on the global bit-line, however, is so much less than that of the original bit-line (only one bypass switch per segment as opposed to a pass transistor per cell) that smaller prechargers/drivers and smaller sense amps can be used. The end result is a net benefit in the power expended to operate the combined system [125, 83].

4.8.5   Further Reading on Cache Reconfiguration

Ranganathan, Adve, and Jouppi proposed reconfigurable caches that can be divided into multiple partitions [189]. Their cache partitioning scheme is similar to the selective cache ways in that it works at the granularity of cache ways. In contrast to the selective cache ways, which only allows two partitions (an enabled and a disabled partition), this proposal allows multiple partitions to be created; up to four partitions can be created in a 4-way set associative cache. Furthermore, cache partitions can be devoted to different functions rather than just being enabled or disabled. The example described in the paper uses one partition as an instruction-reuse cache, i.e., to cache the outcome of frequently appearing instructions. Supporting diverse functionality requires additional address and data busses to accommodate simultaneous access to all the possible partitions. The proposal is focused on performance rather than power and the authors acknowledge that some of their design decisions may actually increase power consumption. However, it is closely related to the low-power proposals discussed above, often resorting to similar solutions for problems such as data accessibility among partitions, replacement, etc.

4.9   PARALLEL SWITCHING-ACTIVITY IN SET-ASSOCIATIVE CACHES

Besides cache resizing which relates to cache capacity, one can attempt to optimize switching on the basis of individual cache accesses (for a fixed capacity). Invariably, the effort to reduce switching activity for an access centers on set-associative or fully associative organizations. There is not much opportunity in reducing switching activity in a straightforward direct-mapped organization, but the prospects for optimizing a naively-designed associative cache are ample: in its power-challenged incarnation the associative cache consumes power linearly to its associativity.

The parallel search in an associative cache is a prime example of parallel switching activity purposed for performance. While it is known beforehand that all but one of the associative ways will fail to produce a hit, all ways are still accessed in parallel for speed.

Figure 4.23 depicts a simplified block diagram of a 4-way set associative cache. Tag and data arrays are shown for four ways. A comparator compares the tags and drives the multiplexor for the data output. Of course, a real implementation could be markedly different in how the tags and data arrays are combined or divided in sub-banks (e.g., the way CACTI

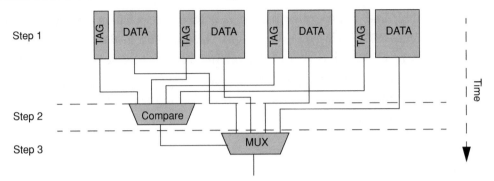

FIGURE 4.23: A power-challenged set-associative cache.

divides up large arrays into sub-arrays). Regardless, the important information conveyed in this figure are the shaded areas of a set-associative cache where switching occurs during an access. In a power-agnostic design, all the cache is shaded: all tag ways and data ways are accessed simultaneously. All the tags of the selected set are matched against the requested address to determine a hit or a miss. Simultaneously, all the data arrays are accessed to provide the data by the time a possible hit is determined.

Clearly we can do better. There is plenty of "excess" switching activity during an access but optimizing it away may cost in performance. The techniques presented here (listed in Table 4.9) aim to significantly reduce power while preserving as much performance as possible.

4.9.1   Phased Cache

A straightforward technique to reduce the full switching activity of a set-associative cache is to defer accessing the data ways until a hit is determined and, then, accessing only the correct data way for the requested data. In other words, as the name suggests, access the cache in phases: first the tags and then (if needed) the data. This technique, appearing as one of the earliest techniques for dynamic power reduction in caches, is discussed in Hasegawa et al. [95] as the implementation of the SH3 cache (Hitachi's low-power embedded RISC processor). Subsequently, it appears in the L2 cache of the Alpha 21264 [87].

The benefit of phasing is a significant reduction in power for the data access which is linear to the miss ratio (no data ways are accessed on misses) and inversely proportional to associativity:

$$P_{\text{data\_new}} = P_{\text{data\_old}} \times (1 - \text{miss\_ratio})/\text{Ways}.$$

The cost in performance is due to the larger latency: the data access no longer can be hidden partially behind the tag access and tag comparison. The performance cost is significant if performance is strongly dependent on latency: for example in non-pipelined L1 caches or

TABLE 4.9: Parallel Switching Activity Optimization in Set-Associative Caches

Technique	What it Does	Examples
Phased access	Finds a tag match first and then accesses the data way	Hitachi SH3 [95] Alpha 21264 [87]
Sequential access	Sequentially accesses ways starting from the way most likely to hit (MRU)	Kessler et al. [133]
Way prediction	Predicts and accesses the cache way that contains the data in parallel to tag match	*MRU prediction,* Inhue et al. [109], *Selective Direct Mapping,* Powel et al. [183], *Multi-MRU,* Zhang et al. [242], Zhu et al. [249]
Way Selection	Deterministically Selects the cache way that contains the data prior to tag match	*Location Cache,* Min et al. [168], *Way Halting,* Zhang et al. [241], *Decaying Bloom filters,* Keramidas et al. [131]

in in-order issue processors. However, if the increased cache latency can be tolerated this is a straightforward and efficient technique to use. Case in point, the Alpha's L2 cache were the additional L2 latency can be easily tolerated by the out-of-order core.

4.9.2   Sequentially Accessed Set-Associative Cache

A sequentially accessed set-associative cache was first proposed by Kessler, Jooss, Lebeck, and Hill [133] as an inexpensive method to implement associativity. In a set-associative organization the sequential access scheme is depicted in Figure 4.25. Initially, only the most likely cache way to produce a hit is probed. In Figure 4.25, the first probe is chosen to be the most recently used (MRU) way. This feature was first proposed for the IBM 370 cache [48]. Additional information is needed to point to the MRU way in each set. However, if the cache features a true-LRU replacement algorithm the MRU information can be readily extracted from the LRU list—incurring of course additional latency. Failing to find the requested data in the MRU way starts a cyclic *sequential* search of the remaining ways.

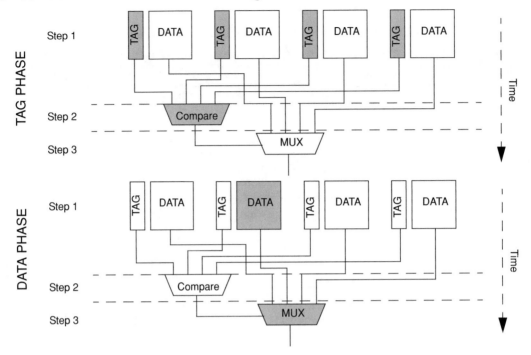

FIGURE 4.24: Phased Cache: Tags first, data later.

Depending on the prediction accuracy, this scheme's performance and power consumption approach those for a small and fast direct-mapped cache. With mispredictions, however, the sequential search can be much more expensive in power consumption than a phased cache and almost certainly slower. Moreover, this scheme suffers considerably on misses since it will consume the maximum energy per access just to find out that it needs to fetch the data from a lower hierarchy level.

☞   *earlier work on pseudo–associativity:* The idea of a sequentially accessed set-associative cache was followed by work on pseudo-associativity which eventually led to way prediction. Early work focused on improving the miss ratio of direct-mapped caches by endowing them with the illusion of associativity [43, 122, 3, 4]. This was driven by performance considerations—and not power which was a secondary concern back then.

Direct-mapped caches are faster than set-associative caches and can be easier to implement and pipeline [99]. Any enhancement that brings their miss ratio closer to that of set-associative caches and at the same time does not compromise their latency can potentially make them top performers. In this direction, work such as the Hash–Rehash cache (Agrawal et al. [4]), the Column associative cache (Agrawal et al. [3]), and the PSA

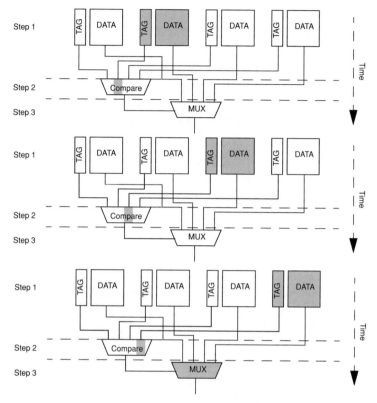

FIGURE 4.25: Sequential access: start from the MRU way and look sequentially.

cache (Calder et al. [43]), convert a direct-mapped cache into a 2-way set-associative cache by mapping conflicting lines to two separate sub-banks.

The lines are still accessed with the benefit of direct-mapping: a single probe checks the tag and reads the data from the most likely location. But if the requested data are not found, an alternative location is also probed. The three proposals differ on how they choose the most likely location for the first access and what they do to improve subsequent accesses, with the PSA proposal fully decoupling location prediction (using a separate prediction structure) from replacement decisions [43]. Although these proposals have substantial differences with respect to power consumption from the corresponding set-associative organizations, no work is available to systematically compare the various options.

4.9.3   Way Prediction

In 1999, Inhue, Ishihara, and Murakami [109] recast the Kessler et al. work on sequentially accessed set-associative caches [133] and the Calder et al. PSA scheme [43], having

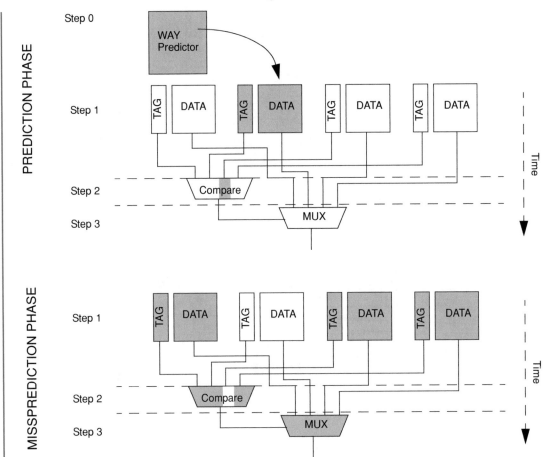

FIGURE 4.26: Way prediction: Choose a way and access tag and data. In case of mismatch, access all the rest of the ways in a second phase for the access.

low-power as the main goal. The general scheme, called *way prediction*, is depicted in Figure 4.26. A separate prediction structure is employed to hold MRU information: for each cache set a bit map points the MRU way with a set bit. This predictor structure is accessed prior to the cache using the requested address (or parts thereof) and provides the prediction of where the requested data are likely to be found.

Initially, only the predicted way (both tag and data arrays) is accessed. Tag comparison determines a hit or a miss for the predicted way. In the case of a miss the remaining ways are accessed in parallel, as in a normal set-associative cache, to determine whether the requested data exist in a non-MRU position. This is a departure from the Kessler et al. scheme, where the remaining ways are searched sequentially and not in parallel. The MRU predictor is updated

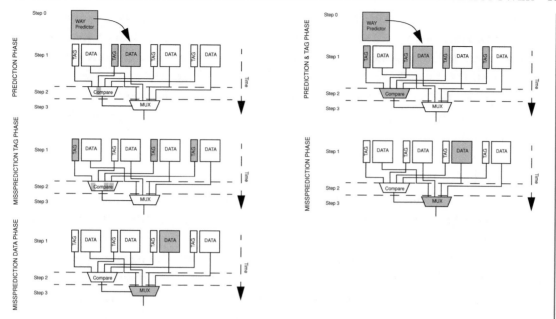

FIGURE 4.27: Hybrid, way prediction + phased access. On the left, way prediction is combined with phased access. The phased access takes place only on mispredictions. The remaining tag ways are accessed in the misprediction tag phase and the correct data way is accessed in the misprediction data phase. On the right, the way prediction and the misprediction tag phase are combined into one. All tag ways are accessed in the first phase along with the predicted data way. In the case of a misprediction, the correct way is accessed in the misprediction phase.

according to where the data are found on hits or with the location of the replaced data on misses. Inhue et al. provide evidence that MRU is sufficiently good prediction in 4-way associative caches to outdo a phased cache in EDP for various SPEC2000 benchmarks.

The power reduction of this scheme is proportional to the prediction accuracy and inversely proportional to the associativity:

$$P_{new} = \text{Accuracy} \times P_{old}/\text{Ways} + (1 - \text{Accuracy}) \times P_{old}.$$

At worse (Accuracy = 0) this scheme cannot be worse in power consumption than a standard cache, but it certainly will be slower (i.e., worse EDP than the base case). Unfortunately, the two factors that determine the power reduction are antagonistic. Higher associativity yields larger benefit on correct predictions, but at the same time lowers the MRU prediction accuracy (resulting in fewer correct predictions).

An improvement in this scheme is to combine it with the phased access approach (Figure 4.27, left diagram). The MRU tag is checked first and in case of a hit the corresponding data way is accessed. If the MRU way does not yield a hit, the remaining ways are also checked, again, in two phases. First, the tags are checked for a hit and then only the correct data way (if any) is accessed for the data. Alternatively, the second step can be combined with the first step into one (Figure 4.27, right diagram). In this scheme, all tag ways are accessed in the first step along with the predicted data way. In the case of a misprediction, the correct data way is known from the tag comparison in the first step and it is accessed in the second (Figure 4.27, right diagram).

4.9.4 Advanced Way-Prediction Mechanisms

The simplicity and success of the way prediction scheme gave rise to more sophisticated prediction mechanisms. Two representative proposals are the selective direct-mapping and the multi-MRU.

Selective direct-mapping: Powell et al. [183], combine selective direct-mapping (SDM) [23] and way prediction in an effort to increase the overall way prediction accuracy. Selective direct-mapping, initially proposed by Batson and Vijaykumar [23], was conceived as an enhancement on the PSA cache. The initial proposal featuring SDM, called *Reactive-Associative Cache*, uses a set-associative organization for tags and a direct-mapped organization for data. Cache lines are mapped in their direct-mapped positions and are only displaced to alternative set-associative positions in the case of conflicts. Cache access resembles a phased access and takes one or two probes to find the data. The first probe, guided by a predictor, accesses either the direct-mapped position or a *predicted* set-associative position. Tag match proceeds in parallel. The second probe only takes place on a misprediction (the data are not found in their direct-mapped position or the way-prediction was incorrect). At that point, the correct position of the data is known from the tag comparison results.

SDM increases the chances of finding the desired data in the first probe. Why is that? In a typical set-associative cache a line can end up anywhere in a set. This is due to the highly dynamic nature of the associative replacement algorithms. Way prediction has the difficult task of predicting the position of each cache line among all the equiprobable associative positions. The problem, of course, worsens as associativity increases. Selective direct-mapping offers a way out by separating conflicting from non-conflicting cache lines.[9] In essence, SDM pins down non-conflicting cache lines to known positions (their corresponding direct-mapped positions) so it is easy to "predict" where they are. In fact, a predictor in SDM only has to supply a prediction on whether a cache line is conflicting or not. In the case it is conflicting, it is in a set-associative

---

[9]This is achieved by using a *conflict list* that captures conflicts close in time. Cache lines enter the conflict list when they are first evicted from the cache, from their direct-mapped positions. Further evictions while in the conflict list give a cache line the "conflicting line" status and form that point onwards it is displaced in a set-associative position rather than placed in its direct mapped position.

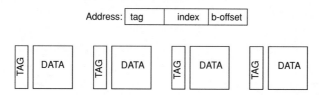

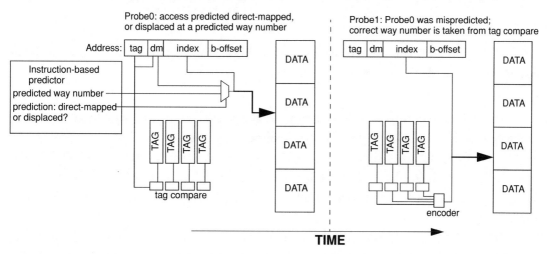

FIGURE 4.28: Reactive-associative cache and selective direct-mapping (with way prediction). The reactive-associative cache maintains the parallel tag banks of a set-associative organization (top diagram) but combines the data banks into one direct mapped bank (bottom diagrams).

position; otherwise it is in its direct-mapped position. If the cache line is conflicting, *further* predicting where it was displaced (way-prediction) is simply an additional bonus. Any other technique such as phased or sequential access could also be used instead. Because way prediction only has to be used on the conflicting, displaced cache lines it is relatively an easier task.

Figure 4.28 shows how SDM works in conjunction with way prediction. The first probe (Figure 4.28, bottom left) is guided by an instruction-based prediction mechanism (see also Figure 4.30 which gives an overview of the prediction schemes). The predictor either directs the first probe to use the *dm* field of the address for the direct-mapped access or supplies a way prediction for the set-associative access.[10] If, according to the tag comparison, the first probe fails, a second probe accesses the data in its correct position, which is encoded from the tag comparison results.

[10]The way-prediction can be supplied by any type of predictor. Powell et al. use a simple history-based predictor indexed by an early approximation of the address. This approximation is derived as the contents of the source address register XOR'ed with any immediate offset found in the corresponding instruction.

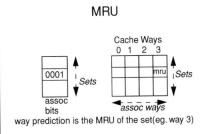

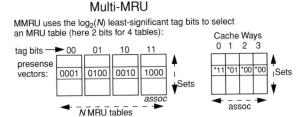

FIGURE 4.29: Multi-MRU way-predictor employs N MRU predictors (typically $N = assoc$) to disambiguate on few least-significant tag bits.

Powell et al. report that SDM combined with way prediction yields significant savings by accessing mostly the direct-mapped or the predicted way. Despite some performance penalty (less than 3%) due to mispredictions, the reduction in EDP is of the order of 64–69% for the 4-way 16KB instruction L1 and data L1, respectively. For their processor models the overall reduction in EDP for this technique is 8%, while with perfect prediction is only 2% better (10%) [183].

Multi-MRU: The *multi-MRU (MMRU)* Zhang et al. proposal [242] (later also appearing in Zhu et al. [249]) is also an extension of the most recently used (*MRU*) way-prediction [43, 48]. MRU simply returns the most recently accessed way of a set as its prediction (Figure 4.29, left diagram) but MMRU allows multiple MRU predictors to *disambiguate* among tags (Figure 4.29, right diagram). All tags in a set having the same low-order bits are tracked by the same MRU table. For example, in Figure 4.29, two tags ending in 00 are tracked by the leftmost MRU table. The prediction is the cache-way of the MRU tag among them (e.g., way 3 in Figure 4.29). In theory, MMRU can disambiguate any number of tag bits, but in practice the technique is limited by the cost of the MRU tables.

It is interesting to note that according to the published results, MMRU is about equal in predictive power to selective direct-mapping when $\log_2$ (associativity) tag bits (i.e., as many MRU tables as the associativity of the cache) are used. In terms of predictive power, SDM aims to place as many lines as it can in their direct-mapped positions and handle the rest with a way-predictor. MMRU tracks all such lines, both those in their direct-mapped position and those in set-associative positions, yielding approximately the same prediction accuracy—an average of 92% first probe hits for 4-way caches [183, 242, 249].

A weakness in all the way prediction techniques mentioned so far is that they do not do well on misses. MRU, MMRU, and SDM incur the maximum latency and energy just to

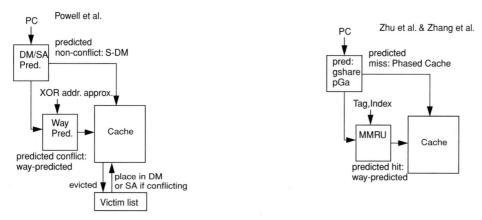

FIGURE 4.30: Overall prediction schemes in [183] (left side) and [242, 249] (right side).

determine that a miss should go to the next level. To address this deficiency, Zhu et al. suggest yet another level of prediction specifically for misses [249]. The resulting scheme is shown in Figure 4.30 (right side). A sophisticated 2-level instruction-based predictor first rules on whether the upcoming access is a miss or a hit. If it is a miss, the cost of accessing the cache is avoided, leaving only the cost of accessing the predictor. If it is a hit, a new prediction (e.g., MMRU or indeed any other way prediction) is used to access only a single way for the data instead of all of the cache.

4.9.5   Way Selection

Way prediction techniques have the disadvantage of the second probe on mispredictions. The second probe costs both in latency and power. In superscalar out-of-order processors, it might also interfere with instruction scheduling, thus incurring additional latency costs [23]. The misprediction rate is also a wild card, meaning that one does not really know how well the prediction mechanism will behave with an untested workload. To address these concerns, a different class of techniques aim to achieve the same goal (accessing a single way) but take prediction out of the picture. We refer collectively to such techniques as *way selection* techniques, alluding to the fact that the way containing the data is deterministically selected prior to the access. Some of the way selection techniques are the Location Cache [168], Way Halting [241], and Decaying Bloom filters [131].

The Location Cache (LC), as its name implies, stores the position of L2 cache lines. The LC sits next to the L1 and being relatively small there is ample time to access it before going to the L2. If, on an L1 miss, the LC supplies a way number, only the appropriate L2 way is accessed. If there is a miss in the LC the L2 is accessed as an ordinary set-associative cache. To

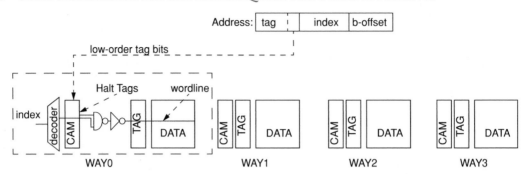

FIGURE 4.31: Way halting: each way is augmented with a CAM array (Halt tags). A Halt tag comprises few low-order bits of the tag. The Halt tags gate the wordlines to the rest of the tag and the data arrays. Adapted from [168].

enhance the chances of hitting in the LC on an L1 miss, sequential prefetching loads the LC with the position of subsequent cache lines.

Way halting is another technique that deterministically accesses only the correct data way. Way halting operates at a single cache level, by halting the parallel access to all the irrelevant ways once a hit and its location are determined in the tag compare. Since tag compare needs to happen very fast for this technique to be successful the authors resort to a partial tag match. Only a few low-order bits of the tag are checked in a CAM structure called *halting tag array*. The CAM structure combines tag access and tag compare in one step (without even needing to decode the index). This makes it fast enough for its outcome to gate the tag and data wordlines driven by the index decoder (Figure 4.31). The access to ways that do not contain the desired data is thus halted by not driving their respective wordlines. Zhang et al. report that way halting can reduce energy in a 4-way cache from 45% to 60% with only a slight area overhead (2%) and without any performance penalty [168].

Finally, Decaying Bloom filters combine a leakage reduction technique (Cache Decay) with a way prediction structure based on Bloom filters [30]. The main intuition for this proposal is that in a decaying cache, dead lines need not be searched. Thus, rather than trying to predict which cache way holds a specific line, the Decaying Bloom filters indicate for each cache way whether the line could be live in it. All the ways that can possibly contain the live line are accessed. In contrast to way-prediction which only accesses a *single* predicted way, this technique may access more than one way but it cannot be wrong: the line is either in the *selected* ways or it is a miss. An additional benefit of using Decay is that Decaying Bloom filters track only the live lines in the ways—dead lines are automatically purged. This considerably increases the accuracy of the information in the Bloom filters. In contrast to way prediction that fumbles on misses as explained above, way selection in general does much better: a miss is detected a lot

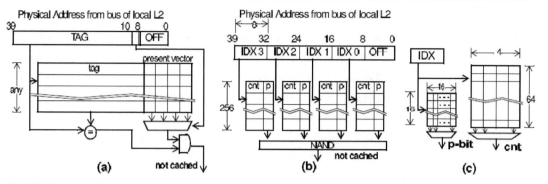

FIGURE 4.32: Bloom filters for Snoops: Include-Jetty and Exclude-Jetty. Reproduced from [171]. Copyright 2001 IEEE.

sooner and the cache access can be avoided altogether. In the Decaying Bloom filters this early detection of misses works very well (when the Bloom filters indicate that there is no live line in *any* of the ways of the cache) giving the technique an advantage over competing way prediction schemes [131].

4.9.6   Coherence Protocols

Cache coherence protocols on bus-based multiprocessors work by having each cache snoop on every bus transaction. This guarantees that if any cache has a copy of the data referenced in the bus transaction, it will also participate in the coherence actions for the data. The very nature of snooping is premised on everyone listening in on the broadcast bus simultaneously, which implies that caches snoop the bus even if they have nothing to do with a particular bus transaction. This brings to mind the parallel access of all the ways of a set-associative cache knowing full well that all but one will miss. However, the distributed nature of cache coherence and the fact that parallel snooping is needed for correctness and not simply for performance make its power optimization an interesting problem.

Moshovos, Memik, Falsafi, and Choudhary proposed the first technique to deal with this problem [171]. Their solution is based on small local filters called *Jettys*. These filters sit between the bus and the L2s and discard most of the unneeded snooping. However, in contrast to way-prediction, mispredictions are not allowed in snoop filtering. *Snoop filtering only works if it can be established with absolute certainty that the snooped address is not cached locally in the L2.* Preventing a snoop to *cached* data is a correctness issue. The Moshovos et al. proposal ensures correctness by using variants of Bloom filters [30] to eliminate snoops that are *guaranteed* to be irrelevant to the contents of the local cache.

Three different approaches are proposed. The first approach is the exclude-Jetty. It is a small tag-cache which is used to indicate *what is not cached* in the L2. It does this by observing misses that go on the bus and remembering when the local cache *does not* have a copy of the data. The exclude-Jetty in this case says "I've seen this before and I am sure its not locally cached."

The include-Jetty, on the other hand, is a Bloom filter and captures a superset of *what is cached* in the local L2. Bloom filters proposed in 1970 by Bloom [30] are hash tables that implement a "*non-membership*" function. Because they can be efficiently implemented in hardware, they are a convenient tool in many situations that require filtering [171, 202, 180, 64].

Each Bloom filter entry is a single bit: a 1 denotes the presence of one or more objects hashed on this entry; a 0 denotes the *absence of any object that can hash on this entry*. A Bloom filter can tell us with certainty when something is *not present*, but it cannot tell what exactly is present because of possible conflicts on its entries. One can arbitrarily lower the probability of conflicts by using multiple hash functions for each inserted object. In this case, an object hashes to multiple Bloom entries and all of them have to be 1 for the object to be present—if any of the entries corresponding to an object are 0, the object is definitely not present.

The include-Jetty can say with certainty that some addresses are not locally cached (if they fail to hit in the Bloom filter Jetty), while other addresses (that hit) *may be* cached locally. For the latter, the snoop proceeds to access the L2 tags to make sure.

Finally, the third approach, the hybrid-Jetty, consults both the include-Jetty and exclude-Jetty for higher efficiency. Moshovos et al. found that 54% of all the snoops miss in the L2 tags in a 4-processor SMP server for the SPLASH-2 benchmark suite. The best Jetty (hybrid-Jetty) eliminates about three quarters (76%) of these snoops yielding analogous power savings. Because the Jettys themselves are tiny compared to the tag arrays of an L2, their operation adds little overhead.

4.10   CACHEABLE SWITCHING ACTIVITY

An important type of switching activity that can be "avoided" to reduce power is repetitive computing activity. In reality, it is not eliminated but converted to caching activity. This is achieved by storing the results of the computation and recognizing when it repeats verbatim producing the same results as before. Instead of re-executing it, a lookup in a cache supplies the results. This can save considerable power if the difference in energy between accessing the cache and re-computing the results is quite large. It is possibly enlightening to consider the cache hierarchy as a recursive application of this concept, only, instead of computation, what is cached in this case is cache activity itself (reads and writes) from a lower—hence, more expensive—level of the hierarchy.

Computation: Repetitive computation when executing a program appears at many levels: at the functional unit (e.g., a multiplier fed by the same inputs), at the instruction level (e.g., the same repeating instruction [208]), at the basic block level (repeating basic blocks such as loop iterations [56]) and at the trace level (groups of instructions in execution order). Such computation, when used with the exact same inputs, produces the same result and therefore can be cached. Many techniques in hardware and even in software—where the compiler discovers the repetition [60, 61]—have been proposed to exploit this property.

A related concept to work reuse is value prediction [156]. Value prediction guesses the outcome of a computation but does not guarantee the correctness of the result. As such, although great for breaking dependence chains by guessing ahead, it requires verification. Full re-execution of the value-predicted computation does not save any switching activity, in fact value prediction adds to the existing switching activity by accessing the prediction structures. For this reason we do not expand further on value prediction.

Cache hierarchy: The cache hierarchy itself, besides a performance optimization, is also a power optimization, in the sense that it steers the majority of accesses to small and power-efficient (lower capacitance) memory structures. To put it another way, the memory hierarchy is a natural way to minimize switching activity in successively larger and more power-hungry caches. A typical cache hierarchy composed of small L1s (instructions and data), and successively larger caches (L2, L3), is intentionally designed so that most accesses are satisfied as close to the processor as possible. The reason why the highest levels of the hierarchy end up with comparably the largest chunk of the power budget is exactly because of this behavior: being more efficient *per access*, they take on the burden of satisfying the *most* accesses.[11] Here, three low-power approaches, exploiting this characteristic of the cache hierarchy are presented: the *filter cache*, the *loop cache*, and the *trace cache*. The last one, the trace cache, combines work reuse (caching the work of instruction decoders) with caching of the instruction L1.

> ☞ *dynamic power in caches*: Dynamic power consumption in caches (but also in other memory structures, e.g., SRAMs, registers, CAMs) depends primarily on two factors: the size of the memory structure (C) and its access activity (A). Size matters, since accessing a physically larger memory requires more power even when the number of accessed bits per access remains constant. This is simply a consequence of the larger decoders and the longer (higher-capacitance) bit/word-lines of larger memories. At the same time, speed is also affected by memory size as a consequence of wire delay being proportional to the square of the wire length. Smaller memory is both faster and more power-efficient. Thus, it is not surprising that caches optimized for speed are also fairly well sized for power. Size optimization, affecting the total capacitance C, in caches is usually done statically via sub-banking, bit-line segmentation (see "Sidebar: Bit-line Segmentation"), etc. CACTI, a popular tool that analytically calculates latency, power, and area for cache organizations, automatically performs such optimizations giving priority to speed optimizations [204].

---

[11] On the other hand going to main memory incurs a significant power penalty at the chip interface because of the chip's I/O drivers and external buses. Fortunately, because of caching, few accesses manage to reach main memory.

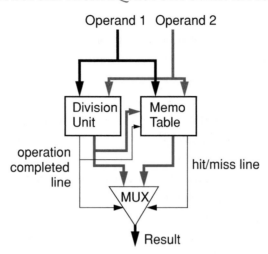

FIGURE 4.33: Operation-level Memoization: The Memo-table in this particular example captures inputs and results from a division unit. When inputs previously seen are detected the result is read from the Memo-table and the computation the division unit ceases.

4.10.1 Work Reuse

Work reuse appears at various granularities. Here, we start at the operation level (finer granularity) and proceed towards instruction, basic block, and trace level (coarser granularities).

Operation level: Citron, Feitelson, and Rudolph proposed *memoization* (or *memoing*) for multicycle floating point operations [56]. Memoization is the act of remembering the result of an operation in relation to its inputs. A memoization cache or Memo-table, as is called by Citron et al. [56], stores the input operands and the result of floating point operations. Upon seeing the same operands the result is retrieved from the Memo-table and is multiplexed onto the output (Figure 4.33). The Memo-table access and the floating point operation start simultaneously. However, accessing the Memo-table is much faster (single-cycle) than performing the actual multi-cycle operation. Since the result is available much earlier this translates into performance benefits but also (by gating the floating point unit before it completes the operation) to power benefits. The power benefits are commensurable to the energy differential between accessing the cache and performing the operation to completion.

Unfortunately, the work of Citron et al. does not contain an evaluation on the power impact of their technique. However, they do give memoization statistics for multimedia applications (Perfect benchmark suite, SPEC FP95, and imaging/DSP applications) which, in conjunction with simple power models for the floating point unit and the memo-tables, can be used to derive power estimates. For their workloads, 59% of integer multiplies, 43% of FP multiplies, and 50% of FP divisions are memoizable and can be "performed" in a single cycle with *small* (32-entry, 4-way set-associative) Memo-tables [56].

Instruction level: Seminal work on dynamic instruction reuse was done by Sodani & Sohi [208]. The observation in their work is that many *dynamically-executed* instructions (or groups of instructions) operate on the *same inputs*. Sodani and Sohi were led to the discovery of this property by examining how execution proceeds in dynamically scheduled superscalar processors. In particular, they noticed that execution in a mispredicted path converges with execution in the correct path resulting in some of the instructions beyond the point of convergence being executed twice, *verbatim*, in the case of a misprediction. Furthermore, the iterative nature of programs in conjunction with the way code is written modularly to operate on different inputs results in significant repetition of the same inputs for the same instructions.

Similarly to the operation memoization, the results of such instructions can be saved and simply *reused* when needed rather than re-executing the computation. Sodani and Sohi claim that in some cases over 50% of the instructions can be reused in this way. Although their work is also focused on performance, the implications of instruction reuse on power consumption can be quite important with such a large reuse rate.

Sodani and Sohi propose three schemes to implement instruction reuse. The first two are simply caches of inputs and results called *Reuse Buffers* (*RB*). One bases its reuse test on input *values*. Upon seeing the same input values for an instruction the result is used. The second simplifies the reuse test and reduces the required storage space per RB entry by relying not on input values but on input register *names*. Reuse of an instruction depends on whether it operates on the same registers as before. RB entries in this case are invalidated when registers are written. In both schemes, the reuse of a load is predicated upon the corresponding memory location *not having been written*. RB entries corresponding to loads are thus invalidated when their address is written. Finally, the third scheme takes into account not only register names but also dependence chains to track the reuse status of such instruction chains. It carries, however, considerable complexity, hence increased power consumption.

Basic block level: Huang and Lilja take reuse one step further and discuss basic block reuse [107]. Their observations concern whole basic blocks for which they found that their inputs and outputs can be quite regular and predictable. Their studies show that for the SPEC95 benchmarks, a vast majority of basic blocks (90%) have few input and output registers (up to four and five, respectively) and only read and write few memory locations (up to four and two, respectively).

Similarly to the RB buffer, a *block history buffer* (*BHB*) stores inputs and outputs of basic blocks and provides reuse at the basic block level. The increased number of inputs that must match for the result to be determinable means that basic block reuse is not as prevalent as instruction reuse. However, when reuse succeeds, not only avoids the execution of the individual instructions in the basic block, but also breaks the dependence chains in it, returning results in a single cycle. In addition to the energy saved by not executing instructions in functional units, considerable energy can be also saved because all the bookkeeping activities in the processor (instruction pointer update, instruction fetch, decode, rename, issue, etc.) during the execution of a basic block are eliminated. Of course, it is much more expensive to access and match entries in the BHB since each entry consists of arrays of values and valid bits [107].

Trace level: similar to the basic block reuse is the trace-level reuse proposed by González, Tubella, and Molina [86]. Traces are groups of consecutive instructions reflecting not their position in the static code layout but their order in *dynamic execution*. A trace may span more than one basic block by allowing executed branches (taken or non-taken) in the middle of the trace. Similarly to basic blocks, a trace too can start with the same inputs, read the same values from memory and produce the same results and side-effects (e.g., memory writes). Trace-level reuse has analogous problems and benefits with basic block reuse, only amplified because the traces can be longer.

4.10.2 Filter Cache

In 1997, Kin, Gupta, and Mangione-Smith proposed one of the first purely architectural techniques to reduce power in cache hierarchies. Called the Filter Cache [142], the idea takes the memory hierarchy characteristic of satisfying accesses in smaller structures to the extreme. The filter cache is a tiny cache (128–256 Bytes) that filters the processor's reference stream in a very power-efficient manner, trading performance for power to yield a better EDP product. The filter cache is inserted between the processor and the L1 which now has a longer latency being farther away from the processor. The original high-performance/higher-consumption configuration with the L1 immediately next to the processor can be restored by simply bypassing the filter cache.

The filter cache satisfies at full speed a significant percentage of the processor's references (about 60% reported in [142]) very economically; but the remaining references that slip to the L1 are slower. The reduced performance due to these slower L1 accesses unavoidably increases program run time. Obviously, the energy benefit of the filter cache must not be outweighed by the extra energy it takes for the longer-running programs, if the overall Energy × Delay of the processor is to be improved. A successful filter cache must strike a delicate balance between its performance (i.e., its hit rate) and its power. A very small filter cache, such as a line buffer—a degenerate case—although quite power efficient, slows down the majority of the accesses that miss in it. This is likely to hurt EDP. On the other hand, immoderately increasing the filter cache's size, or employing full-associativity to increase its hit rate, will seriously diminish its power benefits. A large size increases C, while full associativity increases A since multiple tags must be compared simultaneously.

Designing a successful filter cache is a matter of thorough exploration of the design space to find the points with acceptable performance loss given the power benefits. Kin et al. use their own power models for the cache and parameters for an older 180 nm 3.3 V technology [142]. For MediaBench workloads they observe that for very small filter cache sizes, the increase in hit rate of a fully-associative organization over a direct-mapped organization is not enough to offset its increased power consumption. Thus, for their setup, a fully-associative filter cache is not a good idea. Best results are reported with 128-Byte to 256-Byte direct-mapped filter caches. Taking this work further, one systematically could size the entire memory hierarchy to minimize EDP for specific workloads.

4.10.3 Loop Cache

The counterpart of the filter cache, but for instructions, is the *loop cache* or *loop buffer*. The loop cache is designed to hold small loops commonly found in media and DSP workloads [10, 150, 24]. In contrast to the filter cache which is a full-fledged cache, albeit tiny, the loop cache — or, more accurately, *buffer*— is typically just a piece of SRAM that is software or compiler controlled (a canonical example is found in Lucent's DSP16000 core [10]).

A small loop is loaded in the loop buffer under program control and execution resumes fetching instructions from the loop buffer rather than from the usual fetch path—which might include an instruction L1—until the loop finishes. The loop buffer being a tiny piece of RAM is very efficient in supplying instructions, avoiding the accesses to the much more power consuming instruction L1. Because the loop buffer caches a small block of *consecutive* instructions, no tags and no tag-comparisons are needed for addressing its contents. Instead, only relative addressing from the start of the loop is enough to generate an index to correctly access all the loop instructions in the buffer. Lack of tags and tag comparisons makes the loop buffer far more efficient than a typical cache, even one of the same size.

There are also proposals for fully-automatic loop caches which detect small loops at run-time and install them in the loop cache dynamically [150, 110, 25, 232]. However, such dynamic proposals, although they enhance the generality of the loop cache at the expense of additional hardware, are not critical for the DSP and embedded world where loop buffers have been successfully deployed. This is because in a controlled software environment, the most efficient solution is usually preferable for cost reasons.

In contrast, a fully automatic loop buffer appears in Intel's Core 2 architecture [110]. Intel embeds the loop buffer in the Instruction Queue. A hardware loop detection mechanism, called *Loop Stream Detector* (*LSD*), detects small loops already inside the 18-deep instruction queue. Once a loop is detected, instructions for subsequent loop iterations are streamed from the IQ without any external fetching, until a misprediction on the loop branch is detected. This not only speeds up instruction fetch but at the same time saves considerable energy by not accessing the instruction (or trace cache) and by not decoding the same loop instructions over and over again.

4.10.4 Trace Cache

The concept of storing a trace—a group of consecutive instructions as they appear in *dynamic execution*—and reusing it, was first published by Rotenberg, Smith, and Bennett [193] as a means to increase instruction fetch bandwidth. In this respect it is closely related to the loop cache. However, the trace cache goes further. The idea is to embed branch prediction in instruction fetching and fetch large stretches of instructions despite abrupt changes in the control flow. Although the idea works well for what it was intended for, it found a much more important place as a mechanism to reduce energy consumption for most of the front end of the Pentium-4 processor. This is due to the CISC nature of the IA-32 (x86) instruction set executed by the Pentium-4 [110].

The particularities of a complex instruction set with variable-length instructions such as the IA-32 make it extremely difficult to execute it in a dynamically scheduled superscalar core. Intel's solution is to translate the IA-32 instructions into RISC-like instructions called *uops*. The uops follow the RISC philosophy of fixed length instructions (112 bit long) and of a load-store execution model. IA-32 instructions which can access memory are typically translated into sequences of load-modify-store uops.

The work required in such a front end is tremendous and this is reflected in the large percentage (28%) of the of the total power devoted to the front end. Even before the translation from IA-32 to uop instructions takes place, considerable work is required just to fetch IA-32 *variable-length* (1–15 bytes) instructions, detect multiple prefix bytes, align, etc. Decoding multiple IA-32 instructions per cycle and emitting uops to the rename stage is one of the most power consuming operations in the Pentium-4 processor.

To address this problem Solomon, Mendelson, Orenstien, Almog, and Ronen describe a trace cache that can eliminate the repeated *work* of fetching, decoding, and translating the same instructions over and over again [210]. Called the *Micro-Operation Cache* (µC), the concept was implemented as the trace cache of the Pentium-4. The reason why it works so well in this environment is that traces are created *after* the IA-32 instructions are decoded and translated in uops. Traces are uop sequences and are directly issued as such.

Figure 4.34 shows the concept of the Micro-Operation Cache (adapted from [210]). The µC fill path starts after the instruction decode. A fill buffer is filled with uops until the first branch is encountered. In this respect, the µC cache is more of a basic block history buffer (see BHB, [107]) than a trace cache, but this is not an inherent limitation in the design—it was so chosen just to make it as efficient as possible. Another interesting characteristic of the µC design is that although a hit can be determined in the µC during the first pipeline stage,

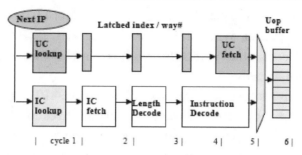

FIGURE 4.34: Micro-Operation Cache (μC) in the P6 architecture. Traces are built as uops are issued after the decode stage. Uop traces are delivered to the issue stage at the same time as the normal front-end path would deliver them. From [210]. Copyright 2001 IEEE.

the uops are not delivered to the issue stage until after 4 more cycles (stages). This ensures that there is no bubble in the pipeline switching back and forth from streaming uops out of the μC to fetching IA-32 instructions from the instruction cache and decoding them.

The benefits for often-repeating traces, of course, are significant. Solomon et al. report that 75% of all instruction decoding (hence, uop translation) is eliminated using a moderately sized micro-operation cache (e.g., 64 sets × 6 associativity × 6 uops/line). This translates to a 10% reduction of the processor's total power for the P6 architecture [210].

The Pentium-4 trace cache is a prime example of a power-saving technique eliminating repetitive and cacheable *computation (decoding)*. But at the same time it is also a cache hierarchy optimization similarly to the loop cache.

4.11   SPECULATIVE ACTIVITY

Speculative switching activity is a high-level type of switching activity relating to speculative execution. Wide superscalar processors need a constant supply of instructions not only to keep multiple functional units busy when this is feasible, but also to make forward progress in the face of costly cache misses. Although there *is* significant instruction level parallelism in many programs, we have come to a point where it is a struggle to maintain an IPC of 1 at the highest frequencies.

Branch prediction is a necessity in this situation. It provides for more independent instructions to keep the functional units busy until the next cache miss. However, even sophisticated branch prediction may not be enough to avoid complete stalls [126]. Prediction, of course, leads to speculation: instructions are executed speculatively until the correct execution path is verified. Besides the actual power consumption overhead of supporting branch prediction and speculative execution (e.g., prediction structures, support for checkpointing, increased run-time state, etc.) there is also the issue of incorrect execution. Incorrect speculative execution that is discarded when the branch is resolved is—for the most part—wasted switching activity. This switching activity cannot be pinpointed to something in particular but it consists of everything that is going on in the processor to execute incorrect instructions.[12] It is therefore orthogonal to all other types of activity discussed in this chapter. It is only characterized as switching activity executing down the wrong path.

A solution for this type of activity was alluded in Section 4.10. Sodani and Sohi observed that a good deal of instruction reuse is due to speculative execution down the wrong path. This is because many times the wrong path and the correct path of execution converge, sometimes quickly, resulting in the same instructions being executed twice: at first following the misspeculation and then again after the branch is resolved. An instruction reuse buffer can capture some of this repetition and reduce the negative impact of incorrect execution, but such a technique has not been researched from a power consumption perspective. Failing to salvage some of the incorrect execution, another *high-level* approach is needed to curb the power impact of incorrect execution.

Pipeline gating: This approach, proposed by Manne, Klauser, and Grunwald, is called *pipeline gating* [161]. The idea is to gate and stall the whole pipeline when the processor treads down very uncertain (execution) paths. Since pipeline gating refrains from executing when confidence in branch prediction is low, it can hardly hurt performance. There are two cases when it does: when execution would eventually turn out to be correct and was stalled, or when *incorrect* execution had a positive effect on the overall performance (e.g., because of prefetching). On the other hand, it can effectively avoid a considerable amount of incorrect execution and save the corresponding power. Saving power without affecting performance is the ideal goal for an effective architectural technique.

The success of pipeline gating depends on how *confidence* in branch prediction is assessed. Two metrics matter for a confidence estimator. First, how many of the mispredicted branches can be detected as low-confidence—this is the *coverage* of the estimator. Second, out of those detected low-confidence branch-predictions how many turn out to be wrong. Since what is of interest here is to detect *wrong* predictions, this is the *"accuracy"* of the estimator.[13] Coverage and accuracy are usually antagonistic in a design. Increasing one diminishes the other. It turns out that it is easier to increase the coverage than the accuracy of an estimator. The estimators proposed by Manne et al. range in coverage from 72% to 88% (for gshare and McFarling combined gshare+bimodal branch predictors) but can hardly reach 37% accuracy in the best case. This shows that even low-confidence predictions are usually—two out of three times—correct.

---

[12]This includes the fetching, decoding, renaming, issuing, and executing of instructions, but of course not the final committing.

[13]For convenience, the terms "coverage" and "accuracy" are used here in the place of the more rigorous terms Specificity and Predictive Value of a Negative Test [161].

Such a low accuracy for the estimator is disheartening for pipeline gating. Most of the time it would stall correct execution. However, this holds for a *single* low-confidence branch.

If more than one low-confidence branch enters the pipeline then the chances of going down the wrong path increase substantially. In fact, for N low-confidence branches and an average estimator accuracy of P (for each), the probability of going down the wrong path (i.e., having at least one misprediction) becomes: $1 - (1 - P)^N$. Conveniently enough, evidence shows that low-confidence predictions *do* tend to cluster together [88]. Pipeline gating is thus engaged with more than one low-confidence branch in the pipeline—the actual number is called *gating threshold*. This makes the coverage of the estimator (detecting many low-confidence branches) more important than its accuracy because it is the number of low-confidence branches in the pipeline that matters—not their accuracy. Manne et al. discuss several possible confidence estimators for the *gshare* and the *McFarling* predictors, including

- *perfect* (oracle) confidence estimation,
- *static* (profiled) estimation allowing the customization of coverage versus accuracy,
- *Miss Distance Counter* (*MCD*) estimator that independently keeps track of prediction correctness,
- for the McFarling predictor an estimator—called *"both strong"*—based on the agreement of the saturating counters of the gshare and bimodal components, and
- finally, for the gshare predictor a simple estimator based on the *distance* of a branch from the last low-confidence branch.

Estimator details are not of much importance here, but rather the fact that different estimators can be designed trading coverage and accuracy. Choosing the *distance* for gshare and *both-strong* for McFarling and with a gating threshold of 2, a significant part of incorrect execution is eliminated without any perceptible impact on performance.

To conclude this approach, one last question that needs to be addressed is the specific pipeline stage to gate. The earlier the pipeline is gated, the more *incorrect* work is saved but also the larger the penalty of stalling *correct* execution. This is not simply a function of the number of pipeline stages before gating. The important factor here is the number of incorrect instructions as we go deeper into the pipeline. Gating at the issue stage hardly saves any extraneous work since very few incorrect instructions make it that deep in the pipeline. In contrast, the initial stages of fetching, decoding, etc. can be full of incorrect-path instructions. With a gating threshold of two or more, the chances of stalling correct execution are miniscule, so it pays to gate as soon as possible (i.e., at the fetch stage).

Selective throttling: Subsequent work by Aragón, J. González and A. González followed a different path. Instead of having a single mechanism to stall execution as in Manne et al.,

Aragón et al. examine a range of throttling mechanisms: fetch throttling, decode throttling, and selection-logic throttling [16]. As throttling is performed deeper in the pipeline, its impact on execution is diminished. Thus, fetch throttling—at the start of the pipeline—is the most aggressive in disrupting execution, starving the whole pipeline from instructions, while decode or selection-logic throttling—deeper in the pipeline—are progressively less aggressive. This is exploited in relation to branch confidence: the lower the confidence of a branch prediction the more aggressively the pipeline is throttled. The overall technique is called *selective throttling*.

Pipeline gating, being an all-or-nothing mechanism, is much more sensitive to the quality of the confidence estimator. This is due to the severe impact on performance when the confidence estimation is wrong. Selective throttling, on the other hand, is better able to balance confidence estimation with performance impact and power savings, yielding a better EDP for representative SPEC 2000 and SPEC 95 benchmarks [16].

4.12   VALUE-DEPENDENT SWITCHING ACTIVITY: BUS ENCODINGS

We conclude our classification of excess switching activity with an activity type appearing at a low level, at the circuit or wire level. It is well known that switching activity in combinatorial and sequential circuits varies with the inputs. For example, an adder exhibits different switching activity depending on the numbers it adds. This raises the possibility of finding a different encoding for the inputs that would lower the switching activity in typical operation. Although it is standard practice in industry to test various encodings of data locally available at ciruit level (e.g., using negated inputs) to see which one minimizes switching, here, we are concerned with higher-level data encodings rooted on architectural properties. As such, most of the published work in the area concerns data encodings for buses.

There are strong incentives to use data encoding on buses. First, buses consume a significant amount of the total processor power so it pays to optimize their power consumption as much as possible. Second, there are only two major factors that drive the power consumption of a bus: the average number of signal transitions on its wires, and the capacitance of its wires. Wire capacitance is mainly a circuit/device level issue and there are solutions to address it at that level; it does not depend on the data carried on the bus (except for *cross talk)*. But the number of signal transitions on the bus does depend on the data carried on the bus and this can be dealt effectively via data encoding.

For off-chip busses, the benefits of encoding are even greater since their capacitive load is orders of magnitude larger than that of internal buses [53]; thus, significant power can be saved during off-chip transmission by encoding alone. In addition, reducing switching activity can be thought of, not only as a way to decrease power consumption, but also as a way to increase the apparent bandwidth of a bus or the I/O pins. In fact, some of the work we discuss below has bandwidth, rather than power, as its main objective for optimization. Our approach for presenting this work is to cluster the proposed techniques into two groups: (i) specialized techniques to reduce switching in address busses—the low hanging fruit in this case—and (ii) more general techniques applicable to both data and address buses.

4.12.1 Address Buses

Early work on bus encoding focused on address busses aiming to exploit their regular sequential and stride behavior. One of the first encoding proposals, although not specifically for reducing power consumption, is Dynamic Base Register Caching by Farrens and Park [75]. They show that high-order address lines exhibit significant temporal and spatial locality and in most cases transmit only redundant information. In their Dynamic Base Register Caching scheme, each address is transmitted in two components: a high-order component (called base) and a low-order component (called offset). The goal is to transmit mainly the low-order components that change frequently, and only rarely the high-order components.

Low-order components are transmitted directly from processor to memory at all cases, while high-order components are stored in a set of base registers located in both the processor and in the memory. When the processor issues a new address, a search is performed in the set of the base registers on the processor side. The goal is to find a register whose context matches the high-order portion of the issued address. The set of registers is organized as a cache. On a hit, the index to the cache entry is sent to memory, instead of the high-order portion itself. When memory receives such an index, it extracts the high-order component of the address from its own set of base registers. To formulate a complete address, the indexed base register and ensuing offset are concatenated.

If the high-order component of an address is not found in one of the processor's base registers, a fault occurs. One of the base registers at the processor side is immediately replaced with the new high-order component. The memory is informed of this fault by the transmission of a reserved index. The update process ends when the processor sends the corresponding base register to memory. The authors examined both fully associative (LRU replacement) and direct-mapped caches for the based register set.

The work of Farrens and Park is about I/O bandwidth rather than power consumption—in 1991 performance was far more important than power. Not surprisingly, not much attention is paid to the fact that what matters in bus switching is whether bus lines change value from one cycle to the next. Thus, even if the high-order component was transmitted all the time, it would not affect power consumption much since it would remain the same for most consecutive addresses. Nevertheless, Farrens and Park exposed an important property of address busses.

Bona fide low-power approaches for address busses were proposed by Owens et al. [176] and Benini et al. [27]. Both schemes are based on the sequentiality of the addresses as they appear on address buses. In the first, Grey-code encoding is used to minimize transitions for sequential values [176]. In the second, a simple but effective scheme, called T0 encoding, is used. The basic idea in T0 is to freeze the bus if the address to be sent is consecutive to the previous address. In such case, a control signal is asserted and the receiver (processor or memory) generates the address locally by incrementing the last address it received.

Working Zone Encoding (WZE) by Musoll, Lang and Cortadella [173] is a retake on the Farrens and Park work but now targeted for low power. The key observation is that the address bus does not behave completely sequentially because accesses to different "zones" are usually interleaved. To solve this problem, they introduce a scheme which keeps track of a few working zones favored by the application. Whenever possible, addresses are expressed as a working zone offset along with an index to the working zone. The offset being small enough can be encoded with one-hot encoding which flips only two bits whenever there is a change in the value transmitted (one-hot encoding requires as many wires as there are values to be encoded, so it is only feasible for small values).

4.12.2 Address and Data Buses

The category of techniques that apply both to data and address buses includes one-hot encoding, bus inversion encoding by Stan and Burleson [212] and its variants (e.g., the combined bus-invert/T0 [28]), INC-XOR encoding [188], and others. The basic bus-inversion coding scheme transfers a data value either in its original form or in its compliment form depending on whose Hamming distance to the previous bus transmission is smaller. According to this scheme, the Hamming distance between the new bus value to be transmitted and the current value is computed. If this is greater than half the number of total bits, then the data value is transmitted in inverted form. An additional "invert" signal is also sent to indicate how the data is to be interpreted at the other end.

Encoding data values is not as straightforward as addresses since data streams are much less regular than address streams. Early work in this area relied on the statistical properties of data streams. The key idea is to use profiling to compute codes such that pairs of values with higher probability of occurrence lead to fewer transitions. Algorithms to analyze profiled data and produce appropriate encoding and decoding functions were investigated by Ramprasad et al. [188] and subsequently by Benini et al. [26]. An adaptive technique which does not require a priori (profiled) knowledge of the statistical properties of the data stream was also proposed in [26]. This technique adapts encoding/decoding to the statistical properties of the data stream on the fly but costs significantly in hardware. We will not expand on these techniques, but the interested reader can find many more details in [26, 188].

Instead of trying to encode data streams via functions, an alternative approach is to utilize dictionary-based solutions. The first approach was proposed by Citron and Rudolph [55]. They describe a technique to encode data on the bus using a table-based approach, called the Bus Expander. The Bus Expander is a cache-like memory structure which is located between a device (processor or memory) and the system bus. According to this scheme, a data item to be transferred over the bus is divided into two parts. The lower part is sent over the bus without modification, while the upper part is inserted in a Look-Up Table (LUT) and its location in the LUT is transmitted instead. In a direct-mapped LUT the location of an item is simply its LUT index. However, in a set-associative, n-way LUT, the location of an item consists of the set number and the way number where it resides. The location is thus derived by dividing the upper part of the data into two fields: the tag and the key. The key is the index to the LUT specifying a set. The items of the set are searched for a matching tag. If a match occurs, the output is the way number where the tag is found. The key, the way number, and the low order bits are assembled and sent over the bus. The Bus Expander on the other side of the bus is responsible for analyzing the transmitted data and recreate the original value. When a value is not found in a LUT, all the Bus Expanders attached to the bus are updated (so future references to this value will not miss). The update leverages a snoopy bus and two control signals—actually, two overloaded data lines—to keep all LUTs consistent. As with the Farrens and Park work [75], this work also focused on increasing effective bus capacity in uniprocessor and multiprocessor systems and not particularly on reducing power consumption.

Instead, this was done by Yang, Gupta, and Zhang who re-evaluated the aforementioned scheme but from a low-power perspective by taking into account the temporal behavior of data values [233]. In their proposed scheme, called Frequent Value Encoding (FVE), the authors show that some values transmitted over data buses tend to recur frequently. A small number of values can account for as much as 32% of the transmissions on the bus [233]. To take advantage of this phenomenon, two identical dictionaries (called *codebooks*) are placed on both ends of the memory channel. The dictionaries can be static or dynamic. Static dictionaries are filled with the most frequent values gathered offline using profiling, while dynamic dictionaries, organized as a linear list to assist LRU replacement, are filled online. To keep the sender side and the receiver side dictionaries consistent, identical LRU replacement is used on both sides. Once a frequent value is inserted in all the dictionaries, it is then transmitted over the bus using one-hot encoding, i.e., a single bus line is asserted to indicate transmission of the corresponding frequent value. One-hot encoding limits the number of frequent values that can be accommodated in the dictionaries to the number of bus wires. An additional control signal is required to disambiguate between one-hot transmissions and non-encoded value transmissions that happen to appear identical on the bus. FVE proved to be quite effective in reducing the data bus switching activity even with very small tables (128 entries).

4.12.3 Further Reading on Data Encoding

A combination of the FVE and the Bus Expander was further evaluated by Basu et al [22]. In this scheme, called Power Protocol, the authors carefully design the update policy of the dictionaries—called Value Caches in [22]—leading to greater reduction in switching activity. Finally, two other techniques are worth mentioning. The first is FVMSBLSB [215] by Shuresh et al., which is another variant of the Frequent Value approach. The difference is that FVMSBLSB stores the MSB portions and the LSB portions of values in separate tables. While encoding MSB/LSB portions alone, the remaining portion of the data are sent un-encoded. Recently, Shuresh et al. [214] proposed a scheme called TUBE which encodes data value segments of varying widths.

4.13   DYNAMIC WORK STEERING

Dynamic work steering refers to the class of techniques in which work is dynamically steered to different components which are alternative implementations for the same function. These components offer a range of power/performance characteristics; for example, one component can be fast and power hungry, while another is slow but power-efficient. One can dynamically choose on which component to perform a function in order to achieve run-time performance or power goals.

Viewed in terms of switching activity (A) and capacitance (C), work steering increases the total capacitance by accommodating multiple implementations of the same function; but, at the same time switching activity is highly focused (dynamically) on one of the alternative implementations. The end result is a highly dynamic modulation of the *effective switched capacitance*.

Work steering can be applied to different levels of granularity: from circuits, to complex microarchitecture units such as the issue logic, or to whole processor cores in a multi-core architecture. Work steering may help to reduce both dynamic and static power. It is orthogonal to the types of excess activity discussed in the previous sections. Orthogonal in the sense that one can derive alternative implementations of the same component, differently optimized for performance and power with respect to a specific type of excess activity and select among them dynamically. In this section, two work steering techniques mentioned previously are reviewed, and a technique at the processor-core level that applies to multi-core architectures is presented.

Circuit level: At the circuit level, *Precomputation* can be considered a rudimentary form of work steering [11]. In this technique, work is steered depending on a simple condition to two optimized precomputation circuits, one for the specific condition and one for its negation.

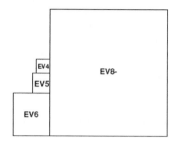

Core	Issue Width	Execution
EV4	2	In-order
EV5	4	In-order
EV6	6	Out-of-Order
EV8-	8	Out-of-Order

Core	Area (mm^2)	Peak Power (Watts)	Typical Power (Watts)
EV4	2.87	4.97	3.73
EV5	5.04	9.83	6.88
EV6	24.5	17.80	10.68
EV8-	236	92.88	46.44

FIGURE 4.35: Estimated values for power and area for the cores used by Kumar et al. [147]. From [147]. Copyright 2003 IEEE.

Section 4.2 contains more details and techniques for handling *Idle-unit activity* involved in this case.

Microarchitecture level: At a higher level, the microarchitecture level, Rochecouste, Pokam, and Seznec proposed a work steering approach for *idle-width activity* [192]. Their proposal is a 4-issue, *width-partitioned microarchitecture* (WPM) processor comprised of two clusters: one normal, 64-bit wide, 2-issue cluster and another, narrow-width, 16-bit, 2-issue cluster. Instructions are steered to the appropriate cluster according to the predicted width of their operands and result. Because this is a statically partitioned microarchitecture, it is complexity-effective, consumes less power, and requires less area than other approaches for idle-width activity. The drawback is that it is only balanced for specific workloads that have an even mix of narrow and wide operands. If the operand width characteristics of the workload differ significantly the performance of the width-partitioned microarchitecture could be compromised.

Analogous approaches at this level could be devised for many of the techniques (and the types of excess activity) discussed in this chapter by offering optimized and "unoptimized" versions of the same structures side by side and dynamically choosing among them according to program and run-time needs.

Processor core level: At this level, Kumar, Farkas, Jouppi, Ranganathan, and Tullsen proposed a multi-core architecture using a variety of cores implementing the same Instruction-Set Architecture (ISA) [147]. As basis for their study they use the Alpha ISA and its implementations, EV4, EV5, EV6 [134], and a single-threaded version of the EV8 (denoted *EV8-*), for which there are plenty of published results for both power and performance.

Figure 4.35 shows the estimated values for the area, peak, and typical power consumption of the four cores examined. To derive these estimates the authors use published results from a variety of sources including, peak to typical power ratios for same-generation Intel processors, and extrapolations from available thermal data. They also assume that the variance in typical power consumption increases in more complex cores due to the wider issue width and increased clock gating. To model power in an architectural simulator executing SPEC2000 benchmarks, Kumar et al. use the activity-based Wattch power models but calibrated accordingly for each core. This is done with the help of scaling and offset factors so as to match the results of the simulator with the estimated peak and typical power consumptions of the cores.

The multi-core architecture in this study is used in a specific way: only one application is run at any one time, i.e., only one core is active. The appropriate core to run the application is chosen to optimize a given objective function (a combination of energy and performance goals). All other cores are powered down expending neither dynamic nor leakage power. Because there is a cost to switch an application from one core to another, the granularity of switching is kept at the OS scheduling quantum (task switching). This is convenient for two reasons. First, the operating system can orchestrate the core switching. Second, saving and restoring the processor state happens by default at the scheduling interval so it does not represent an additional overhead for core switching.[14] Alternatively, choosing on which core to run an application could be performed (even statically) at the granularity of an entire application, but this would preclude adaptation to the needs of individual *program phases* of the application.

Kumar et al. show that both power and performance vary considerably depending on program phase. On the same core—as expected—performance varies from phase to phase. But more importantly, the relative performance difference among phases depends on which core executes the application. For instance, running on EV4, the performance difference among phases might not be that great; in contrast, it can vary widely on EV8-. This makes the *relative performance among cores* vary according to application phase: in some phases EV8-performance is much higher than in other cores; in other phases the performance difference is hardly noticeable.

Things are more interesting when, in addition to performance, energy is taken into account. Tracking Energy × Delay across different phases on the same core shows that the relative difference for this metric *across cores* also varies with phase. In addition, *the ordering of the cores based on this metric is frequently upset*! This means that sometimes EV4 can have a better *EDP* than EV8- and vice versa! This of course is a strong incentive for core switching on a phase granularity. Note, however, that no interval-based approach, even with oracle knowledge, can guarantee the *global* optimization of EDP or ED^2P. See "Sidebar: Pitfalls in optimizing EDP."

---

[14]L1 caches are local to each core, so only the contents of the shared L2 cache are preserved across a core switch.

Kumar et al. examine the optimization of two different objective functions, one minimizing Energy and the other EDP, assuming either oracle heuristics for core switching or realistic dynamic switching heuristics. Under the oracle heuristics the operating system simply chooses the core for each time quantum that optimizes the objective function. Switching overhead is not taken into account.

Lack of oracle knowledge dictates that realistic heuristics must discover for themselves which core optimizes the objective function. Every hundred time quanta, cores are tested for five consecutive quanta to assess their effectiveness in optimizing the objective function. Switching overhead is accurately modeled in the simulations. Thus, the effectiveness of a core on the first of the five test quanta is ignored to discount the effects of cold caches, cold TLBs, and cold branch predictors.

The four realistic heuristics studied by Kumar et al. differ on which cores are selected for testing:

- *neighbor*: a neighbor in the performance continuum to the core that is running is selected at random;

- *neighbor-global*: similar to the above but the selected core is the one that best optimizes the objective function in the application's execution thus far;

- *random*: any core is selected at random;

- *all*: all cores are tested.

In all cases, the authors report substantial gains in energy or EDP. For the oracle heuristics, optimizing for energy yields up to 60% energy reduction with a 5% performance loss [147]. EDP is reduced by up to 63%. Translating these results into ED^2P shows that core switching can even outperform chip-wide DVFS—which can, at best, only break even on the ED^2P metric.

Results are also exceptionally good for the realistic heuristics with all switching overhead accounted. EDP for three of the four heuristics (*neighbor*, *neighbor-global*, and *random*) is within 90% of the oracle heuristics. The "all" heuristic tends to fall behind because of its greater testing overhead.

Overall, core switching is an important work steering technique that can be used not only for reducing power consumption, but also for reducing power density (by migrating activity from hot and busy cores to idle and cold cores). In this last incarnation, the technique is known as *activity migration* [97].

☞ *pitfalls in optimizing EDP*: Kumar et al. noticed a peculiar behavior in their interval-based approach to optimize EDP. Even when choosing the optimal EDP per interval, with

TABLE 4.10: The Interval-based Core-switching Approach Selects P1 in the First Interval and P2 in the Second with Oracle Knowledge. The Resulting EDP (18) is Worse Than Either P1 or P2 Running Both Intervals (EDP of 16.2 and 16.4, Respectively). Adapted from [198].

	P1			P2			Best
Interval	Energy	Time	EDP	Energy	Time	EDP	EDP Core
1	1	4	4	2	2.1	4.2	P1
2	1	4.1	4.1	2	2	4	P2
Overall	2	8.1	16.2	4	4.1	16.4	$(1+2) \times (4+2)$ $= 18$

oracle knowledge, the overall EDP for the whole application may turn out to be far from optimal. Sazeides, Kumar, Tullsen, and Constantinou investigated the matter further and realized that *it is not possible to guarantee EDP—or for that matter ED²P—optimization with any interval-based approach where local, per-interval, decisions are taken* [198]. The root of the problem is that choosing the smallest energy–delay *product* (or energy–delay² product) regardless of the magnitude of its factors may result in globally sub-optimal decisions. To illustrate the problem the core switching example in Table 4.10 shows how the interval-based approach, while choosing the best core (best EDP) per-phase, fails to optimize the global EDP:

More formally, optimizing EDP for an application is equivalent to solving: MIN $(E \times t)$, where E is the energy spent during the application execution time t.

Dividing application execution into n intervals, the problem becomes

$$\text{MIN} \left(\left(\sum_{i=1}^{n} E_i \right) \times \left(\sum_{i=1}^{n} t_i \right) \right),$$

where E_i and t_i are the energy and duration, respectively, of interval i.

If there are *several* choices in interval i for the pair E_i and t_i (for example, *several* core switching choices), then choosing option j in interval i is denoted by E_i^{ji} and t_i^{ji}.

In this case, optimizing EDP *individually per interval* is equivalent to finding the j for each i such as each *product term of the sum* is minimal:

$$\sum_{i=1}^{n} \text{MIN}\left(E_i^{ji} \times t_i^{ji} \right).$$

The sum is minimized, but *this is not equivalent* to finding the optimal global EDP. This, in contrast, is done by selecting the j for each i with the intent of minimizing the following *product of sums*:

$$\text{MIN}\left(\left(\sum_{i=1}^{n} E_i^{ji}\right) \times \left(\sum_{i=1}^{n} t_i^{ji}\right)\right).$$

It is possible, therefore, for the two formulae above to have significantly different solutions. The same can be shown for ED^{2}P [198]. In contrast, energy can be safely optimized with interval-based approaches using just local decisions [198].

CHAPTER 5

Managing Static (Leakage) Power

Static power consumption has grown to a significant portion of total power consumption in recent years. In CMOS technology, static power consumption is due to the imperfect nature of transistors which "leak" current—thereby constantly consuming power—even when they are not switching. The advent of this form of static power, called *leakage power*, was forecasted early on [32, 136], giving architects the opportunity to propose techniques to address it. Such techniques are the focus of this chapter.

Considerable work to reduce leakage power consumption is taking place at the process level [31]. In fact, process solutions such as the *high-k dielectric materials* in Intel's 45 nm process technology, are already employed. Addressing the problem at the architectural level is, however, indispensable because architectural techniques can be used orthogonally to process technology solutions. The importance of architectural techniques is magnified by the *exponential* dependence of leakage power to various operating parameters such as supply voltage (V_{dd}), temperature (T), and threshold voltage (V_T). Exponential dependence implies that a leakage-reduction solution that works well at some specific operating conditions may not be enough— the problem is bound to reappear with the same intensity as before but at higher temperatures or lower voltages.

Undeniably, the most fruitful ground for developing leakage-reduction techniques at the architectural level has been the cache hierarchy. The large number of transistors in the on-chip memory largely justifies the effort (or obsession) even though these transistors are not the most "leaky"—that distinction goes to the high-speed logic transistors [41]. In addition, the regularity of design and the access properties of the memory system have made it an excellent target for developing high-level policies to fight leakage. Most of the architectural techniques presented in this chapter, therefore, target caches or memory structures.

Chapter structure: The presentation of techniques in this chapter is structured according to the type of low-level leakage-reduction mechanism employed (Table 5.1). Architectural techniques inherit similar characteristics according to the physical quantity that is manipulated by their low-level, leakage-reduction mechanism. Here, we concentrate on three major low-level mechanisms (shown in Table 5.1). The first two, the *stacking effect* and the

TABLE 5.1: Structure of the Leakage Reduction Tehniques in this Chapter.

Low Level Mechanism	High-Level Techniques	Characteristics	Section
Stacking effect and gated V_{dd}: sleep transistor cuts off power	Dynamically resized cache (DRI) [239], cache decay [127], adaptive mode control (AMC) [250], functional unit decay [105]	Non-state-preserving (state-destroying) Significant leakage reduction Power-up latency: 10's of cycles	Section 5.2
Drowsy effect: scales supply voltage to reduce leakage	Drowsy caches [77, 137], drowsy instruction caches [138, 139], hybrid approaches (decay + drowsy) [164], temperature-adaptive approaches [129], compiler approaches & hybrids [246]	State-preserving Medium leakage reduction Power-up latency: <10 cycles	Section 5.3
Threshold voltage (V_T) manipulation:	Dynamic Combined V_{dd} (e.g., DVFS) and V_T (e.g., Adaptive Body Biasing—ABB) scaling [163, 231, 70] Static MTCMOS Functional Units [69], Asymmetric Memory Cells [17, 18]	Significant leakage reduction	Section 5.4

drowsy mode, manipulate voltage across transistor terminals (source and drain). This affects the magnitude of leakage reduction, the latency in switching leakage modes, and the ability to retain state in the low-leakage mode. The third class of low-level mechanisms manipulates the transistor threshold voltage (V_T) which can dramatically decrease leakage but at the cost of reduced device speed.

It is important to note here that the techniques presented in this chapter address a specific type of leakage, called *subthreshold leakage*. Another type of leakage, called *gate oxide leakage*, is not addressed architecturally but rather at the process level. To gain a better understanding of the structure of this chapter as well as the difference in the two types of leakage, the following section (Section 5.1) delves into the underlying mechanics of leakage.

5.1   A QUICK PRIMER ON LEAKAGE POWER

Static power is so called because it is consumed by every transistor even when no active switching is taking place. In older technologies (e.g., NMOS, TTL, ECL, etc.) it is an inherent problem, because a path from V_{dd} to ground is open even when transistors are not switching. With the advent of CMOS, static power became less of a concern because the *Complementary* gate design prevents open paths from V_{dd} to ground.

Unfortunately, static power resurfaced in CMOS in the form of *leakage power*. In the latest process generations leakage power increases exponentially, principally because of reductions in the threshold voltage. Leakage power increased to levels never seen before in CMOS—levels comparable to the dynamic (switching) power consumption—when technology scaling entered the deep-submicron territory in feature size (<180 nm). Currently, 20–40% of the total power consumption is attributed to leakage power.

CMOS static power arises due to *leakage currents*. The total leakage current (I_{leak}) times the supply voltage gives the static power consumption, P_{leak}:

$$P_{leak} = V \times I_{leak}.$$

Leakage currents are a manifestation of the true *analog* nature of transistors, as opposed to our idealized view of them as perfect digital switches. The state of a transistor (on or off) is controlled by the voltage on its gate terminal. If this voltage is above the *threshold voltage* (V_T) the channel beneath the gate conducts, allowing current in the on state (I_{on}) to flow from the source (V_{dd}) to the drain (GND, ground). In the opposite case (gate voltage below V_T), we like to think that the transistor is off (perfect insulator). But in reality transistors leak: *leakage currents* flow even in their off state. This is evident in the *I–V* curve where current flows even below the threshold voltage where the device is supposed to be "off."

The current that flows from source to drain when the transistor is off is called *sub-threshold* leakage. But that is not all. There are five more types of leakage: reverse-biased-junction

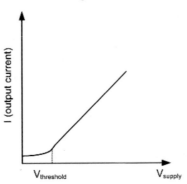

FIGURE 5.1: Example of an "*I–V*" curve for a semiconductor diode (introduced in Chapter 1). Although we informally treat semiconductors as switches, their non-ideal analog behavior leads to leakage currents and other effects.

leakage, gate-induced-drain leakage, gate-oxide leakage, gate-current leakage, and punch-through leakage. The sub-threshold leakage and gate-oxide leakage dominate the total leakage current in devices. Both increase exponentially with each new technology generation with the gate-oxide leakage significantly outpacing the sub-threshold leakage.

In sub-micron technologies, subthreshold and gate leakage is the *cost* we have to pay for the increased speed afforded by scaling. Supply voltage scaling attempts to curb an increase in dynamic power. Unfortunately, this strategy also leads to an enormous increase in the subthreshold and gate leakage problem. This explains why static power has been gaining on dynamic power as a percentage of the total power consumption with every process generation.

5.1.1   Subthreshold Leakage

Subthreshold leakage increases with technology scaling due to V_{dd} scaling. The supply voltage (V_{dd}) is scaled along with other physical quantities to reduce dynamic power consumption. Scaling solely the supply voltage, however, *increases* the *delay* (switching speed) of the transistor. This is because the delay is proportional to the inverse of the current that flows in the *on* state—the I_{on} current (as in the *I–V* curve of Figure 5.1):

$$\text{Delay} \propto \frac{1}{I_{on}} \propto \frac{V_{dd}}{(V_{dd} - V_T)^a}.$$

This current, I_{on}, is a function of the supply voltage and the *difference* between the supply voltage and the threshold voltage (V_T). The factor α is a technology-dependent factor taking values greater than 1 (between 1.2 and 1.6 for recent technologies) [195]. Since V_{dd} is lowered in order to maintain the speed increase from scaling, the only course of action is to also lower the threshold voltage. Herein lies the problem: *subthreshold leakage increases exponentially with lower threshold voltage.*

To understand the basic mechanisms for leakage reduction we have to take a closer look at the formulas describing leakage current. We base our discussion on the Berkeley Predictive Model (BSIM3V3.2) formula for subthreshold leakage [143] (which is also the starting point for the simplified Butts and Sohi models [41] discussed in Chapter 2). The formula describing the subthreshold leakage current, I_{Dsub}, is:

$$I_{Dsub} = I_{s0} \left(1 - e^{\frac{-V_{ds}}{v_t}}\right) e^{\frac{V_{gs}-V_T-V_{off}}{n \cdot v_t}}.$$

Here, V_{ds} is the voltage bias across the drain and the source and V_{gs} is the voltage bias across the gate and source terminal. V_{off} is an empirically determined BSIM model parameter and v_t ($v_t = kT/q$) is a physical parameter called *thermal voltage*[1] which is proportional to the temperature, T. The term n encapsulates various device constants, while the term I_{s0} depends on the transistor geometry (in particular, the aspect ratio of the transistor, W/L).

Immediately, this equation shows the dependence of leakage to W/L, and its *exponential* dependence to V_{ds}, V_{gs}, V_T, and T.

- *W/L, transistor geometry*: Leakage grows with the aspect ratio of a transistor and with its size. Butts and Sohi use simplified models that encapsulate transistor geometry in the k_{design} parameter. They point out that very small transistors such as those found in SRAMs can leak much less than sized-for-performance logic gate transistors. Transistor sizing is primarily a circuit-level concern and it will not preoccupy us at the architecture level.

- V_{ds}, *voltage differential between the drain and the source*: This is probably the most important parameter concerning the architectural techniques developed for leakage. Two important leakage-control techniques that are based on reducing V_{ds} are *the transistor stacking technique*[2] and the *drowsy technique*—a.k.a. dynamic voltage scaling (DVS) for leakage [77]. Both these techniques rely on the $(1 - e^{(-V_{ds}/V_t)})$ factor of the subthreshold leakage equation. This factor is approximately 1 with a large V_{ds} (i.e., $V_{ds} = V_{dd}$ and $V_{dd} \gg v_t$) but falls off rapidly as V_{ds} is reduced. Architectural techniques based on transistor stacking—in particular, a stacking technique called *gated V_{dd}* [184]—and on the drowsy technique form the bulk of the work described in this chapter. The former are presented in Section 5.2 and the latter in Section 5.3.

---

[1]For the thermal voltage equation, k is Boltzmann's constant and q is the magnitude of the electron's charge. At room temperature ($T = 300$ K), the thermal voltage is about 26 mV.

[2]The stacking effect itself is also partially due to a change in the V_T. This chance is dynamic and is caused by a slight reverse bias induced by the top (off) transistor on the bottom (off) transistor.

- V_{gs}, *voltage differential between the gate and source*: Regarding subthreshold leakage for devices in their normal "off" state, this factor can be set to zero, so it is not a concern. Butts and Sohi use this assumption to arrive at their simplified leakage model [41]. However, V_{gs} plays a significant role in the gate-oxide leakage discussed in Section 5.1.2.

- V_T, *threshold voltage*: The threshold voltage—the voltage level that switches on the transistor—significantly affects the magnitude of the leakage current in the *off* state. The exponential dependence of subthreshold leakage on $(V_T)^{-1}$ is evident in the last factor of the BSIM3 formula: the smaller the V_T, the higher is the leakage. Raising the threshold voltage reduces the subthreshold leakage but compromises switching speed.

  Many circuit-level techniques, e.g., MTCMOS, reverse body bias (RBB) and larger-than-V_{dd} forward body bias [13, 174, 14, 222], have been developed to provide a choice of threshold voltages. These techniques provide multiple threshold voltages at the process level (for example, MTCMOS offers high-V_T and low-V_T devices) or vary the threshold voltage dynamically by applying bias voltages on the semiconductor body (e.g., RBB and larger-than-V_{dd} FBB). Architectural techniques based on manipulating the threshold voltage are presented in Section 5.4.

- T, *temperature*: Last but not the least, subthreshold leakage exponentially depends on temperature, T, via the thermal voltage term v_t. This is actually a dangerous dependence since it can set off a phenomenon called *thermal runaway*. If leakage power—or for that matter any other source of power consumption—causes an increase in temperature, the thermal voltage v_t also increases linearly to temperature. This leads, in turn, to an exponential increase in leakage, which further increases temperature. This vicious circle of temperature and leakage increase can be so severe as to seriously damage the semiconductor. The solution is to keep the temperature below some critical threshold so that thermal runaway cannot happen. Cooling techniques, combined with accurate thermal monitoring, are used for this purpose.[3]

  Architecturally, the dependence of leakage to temperature is quite interesting. This is because at low temperatures it might not be so important to engage architectural techniques that could hurt performance with little payoff. As temperature rises and leakage power becomes the dominant component of power consumption (and hence *heat generation*) architectural techniques that can curb leakage become much more appealing. One such example is presented in Section 5.3.4.

---

[3]Unfortunately, the subject of thermal management, despite its importance, is too extensive to receive other than superficial coverage in the space of this book. Here, it is only mentioned briefly with respect to leakage (Section 5.3.4).

5.1.2   Gate Leakage

Gate leakage (also known as gate-*oxide* leakage) is a major concern because of its tremendous rate of increase. It grew 100-fold from the 130 nm technology (2001) to the 90 nm technology (2003) [31]. Major semiconductor companies are switching to "*high-k*" dielectrics in their process technologies to alleviate this problem [31].

Gate leakage occurs due to direct tunneling of electrons through the gate insulator—commonly silicon dioxide, SiO_2—that separates the gate terminal from the transistor channel. The thickness, T_{ox}, of the gate SiO_2 insulator must also be scaled along with other dimensions of the transistor to allow the gate's electric field to effectively control the conductance of the channel. The problem is that when the gate insulator becomes very thin, quantum mechanics allow electrons to *tunnel* across. When the insulating layer is thick, the probability of tunneling across it is virtually non-existent. As the insulating layer becomes thinner, tunneling becomes stronger. Gate-oxide thickness has scaled from 100 nm (1000 Å) to just 1.2 nm (12 Å) in 90 nm and 65 nm technologies. This corresponds to a thickness of just 4–5 atoms [50, 31]! The result is an uncontrollable, exponential increase in gate leakage.

Gate leakage is somewhat dependent on temperature but strongly dependent on the insulator thickness and the gate-to-source (V_{gs}) or gate-to-drain (V_{gd}) biases seen by the device. Without the V_{gs} or V_{gd} biases, the necessary electric field to cause the electrons to tunnel across the gate is absent. Since the supply voltage (V_{dd}) determines the magnitude of V_{gs} and V_{gd}, scaling V_{dd} reduces gate leakage. There is also a weaker dependence of gate leakage on V_{ds}—the voltage across the drain and source—that ties gate leakage to the state of a circuit [190].

The most promising remedy for gate leakage, and the one that is currently in use in the latest generation 45 nm technologies, is to insulate the gate using *high-k* dielectric materials instead of the more common SiO_2 oxide material.[4] A thicker insulating layer of a high-*k* material can be as good as a thin layer of a low-*k* material. The increased thickness significantly reduces the tunneling effect but at the same time does not compromise the ability of the gate to control the channel. In other words, *performance* is not compromised.

Architecturally, gate leakage has not been given the same attention as subthreshold leakage. For the most part, it is considered as an additional leakage component and the hope is that process-level solutions will address the problem. The HotLeakage simulator, mentioned in Chapter 2, takes gate leakage into account, thus giving a more accurate picture for the benefits of various techniques that target subthreshold leakage.

---

[4] A high dielectric constant, *k*, means that these materials concentrate the electric field better. When used as insulators between the plates of a capacitor, a high dielectric constant yields higher capacitance for the same insulator thickness or, alternatively, the same capacitance but with a thicker insulator layer.

5.2   ARCHITECTURAL TECHNIQUES USING THE STACKING EFFECT

Transistor stacking refers to the technique of stacking *off* transistors source to drain [240]. Stacked off transistors, especially if any of them are high-V_T devices, significantly restrict the leakage current flowing to ground. This is because the voltage differential between the drain and source of the stacked transistors is less than V_{dd}. In addition, there is a change in the V_T of the bottom transistor that also helps in reducing the leakage current. A popular stacking technique is the gated-V_{dd} (or gated-V_{ss}) technique developed by Powell et al. for memory cells [184].

The stacking effect, and in particular its *gated V_{dd}* incarnation, has been successfully employed in many architectural techniques, such as the DRI I-cache [184] and cache decay [127] among others. The initial target for these techniques has been the cache hierarchy. The basic strategy is to *turn off* unused parts of the cache to reduce leakage. These techniques are collectively known as *non-state-preserving* (or *state-destroying*). This is because the underlying mechanism destroys all state by cutting off the power supply to the target circuit. Here, we briefly present the gated V_{dd} mechanism and proceed with the architectural techniques that use it.

Gated-V_{dd}: Powell, Yang, Falsafi, Roy, and Vijaykumar proposed the circuit-level mechanism to reduce leakage, called gated V_{dd} [184]. The technique is well suited for use with six-transistor SRAM cells but can also be used in other arbitrary logic circuits.

Gated-V_{dd} is a transistor stacking technique. Figure 5.2 shows a traditional six-transistor SRAM cell and its gated-V_{dd} counterpart. The difference is the *sleep* transistor that gates the ground. In normal operation, the sleep transistor is *on*. Turning this transistor *off* disconnects the SRAM cell from the power supply. While leakage currents in the "off" transistors of the cell are virtually eliminated, the "on" transistors lose the ability to draw current from the power rail since the path to ground is cut off. This means that the feedback loop of the memory cell cannot

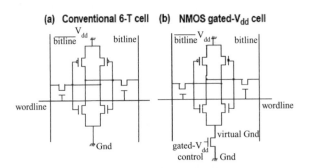

FIGURE 5.2: Gated V_{dd} six-transistor SRAM cell. Reproduced from [239]. Copyright 2001 IEEE.

maintain the charge in its internal nodes. The cell quickly loses its stored value going into a limbo state. Restoring the power supply (turning *on* the sleep transistor) allows the internal nodes of the cell to recharge, but they take on a random logic state.

5.2.1   Dynamically Resized (DRI) Cache

In tandem with the gated-V_{dd} mechanism, Powell et al. [184] and Yang et al. [239] proposed an architectural technique to reduce leakage in instruction caches. This technique, referred to as the *Dynamically Resized (DRI)* instruction cache, is the first of its kind—an architectural technique to save leakage power.

The idea is to resize the instruction cache to fit just the working set of the code that is currently running, turn off the rest of the cache (using gated-V_{dd} sleep transistors), and save the corresponding leakage power. The instruction cache is the obvious initial target for such techniques because of the working set properties of code. Typically, the working set for code exhibits high temporal locality. Consider, for example, loop behavior where for long stretches of time only a fixed, well-defined, part of the *I*-cache is accessed.

The design for a *direct-mapped* DRI cache is shown in Figure 5.3. The design can be easily extended to a set-associative organization. The cache is resized *in its number of sets* by changing the number of index bits with the help of a "*size mask*." The mask disables any number of high-order index bits from left to right. The number of active sets in the cache is *halved* each time an index bit is disabled. To accommodate a varying index size, the tags are extended to also store the maximum number of index bits that could be disabled. Thus, the tags increase in size by the corresponding amount.

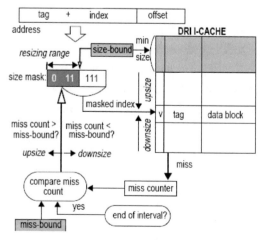

FIGURE 5.3: DRI cache. Reproduced from [239]. Copyright 2001 IEEE.

The downside of resizing the cache in the number of its sets is that already resident cache lines become inaccessible. The oversized tags prevent erroneous matches and make flushing the remaining part of the cache unnecessary. The miss rate, however, is still affected with every resizing since everything is lost in the turned-off part.

The DRI approach to resizing is in stark contrast to the resizing approaches for *dynamic power* covered in Chapter 4, Section 4.8. Techniques such as Selective Cache Ways [8], the Accounting Cache [68, 9], or Miss Tag Resizing [243], resize the data cache to reduce dynamic power by disabling associative ways, i.e., changing the cache associativity. In fact, the Miss Tag Resizing technique [243] also uses the gated-V_{dd} mechanism to completely turn off cache lines and save leakage power along with the dynamic power. One of the benefits of these resizing approaches is that no change in the indexing of the cache is needed, meaning that data already resident remain accessible. Although these techniques could very well be adapted for static power, Powell et al. took the approach of resizing the cache in the number of its sets. Their reasoning is that resizing in associativity is not really necessary for instruction caches because it would preclude direct-mapped caches and would affect both capacity and conflict misses.

Losing the ability to access resident lines might not be as disastrous for instruction caches as for data caches. This is because changes in the working set of code tend to be more abrupt than the corresponding changes for data—execution simply moves to another part of the code, scrapping the previous working set. In addition, the read-only nature of code eliminates consistency and coherency problems stemming from turning off or "misplacing" cache lines in the cache with the new indexing.[5]

Resizing policy: The policy proposed to resize the DRI I-cache is based on monitoring the miss rate. Misses are counted within a fixed time interval (on the order of a few thousands of cycles). At the end of the interval a resizing decision is made. The decision compares the measured number of misses to a user-defined preset "miss bound." If the cache does not perform up to expectations (measured misses > miss bound) the number of sets is increased; otherwise the cache is further downsized. A user-defined "size bound" prevents downsizing of the index beyond some point. This is a safety mechanism to prevent overzealous downsizing.

The *size* bound prevents pathological oscillations between two sizes. This can happen when the miss rate exceeds the bound for the smaller size but is well under the bound with the next larger size. Finally, a parameter, called *divisibility* of the cache controls how many index bits to enable or disable at a time—i.e., it is the divisor (2, 4, 8, ...) or multiplicand for resizing the cache.

Although this policy dynamically resizes the cache under the miss bound constraint, critical parameters such as the size bound, the miss bound itself, and the divisibility factor

---

[5]A notable exception for read-only code is Intel's IA-32 ISA which allows self-modifying code.

are user defined and set individually per program. Thus, the policy stops short of being fully adaptive.

Yang et al. examine a 64KB DRI I-cache with SPEC95 [239]. With hand-selected parameters per benchmark, the DRI cache can save up to 64% of the leakage power incurring a performance penalty of 4% [239].

5.2.2   Cache Decay

Following the DRI cache, the idea of using the gated-V_{dd} technique to turn off cache lines was applied to data caches but at a much finer grain. The technique, called cache decay, turns a cache line off if a pre-set number of cycles have elapsed since its last access [127].[6] However, turning off a cache line that is in active use incurs extra dynamic power re-fetching the line from lower cache levels. Therefore, a central goal of cache decay is to accurately predict when a cache line is no longer useful, or—as it is more commonly known—when a cache line is *dead*.

Cache decay is based on the fundamental *generational behavior* of cache lines. This behavior was first discovered by Wood, Hill, and Kessler in their effort to support faster trace-driven cache simulation [225]. At the time when the Wood et al. paper appeared, it was prohibitive to use a long memory trace in a cache simulator—the limiting factor, of course, being the capabilities of the machines of the time. One solution to this limitation is to skip long stretches of the trace and simulate only some patches of the trace—or *trace samples* as they are known. But what would the state of the (simulated) cache be after skipping a long stretch of the trace? What cache lines would it contain?

The reason why the state of the cache is important is because the miss rate of *unknown* references depends on it. Unknown references are the references that have not been encountered before in the trace. It is unclear if they are actual cold misses for the cache or simply seen for the first time because of part of the trace is skipped. Contrary to intuition—and much to their surprise at the time—Wood et al. found that the miss rate of unknown references is *much higher* (e.g., 0.4 rather than 0.02) than the steady-state miss rate of the cache. How can this be? The answer lies in the generational behavior of the cache lines.

Generational behavior. The very high miss rate for the unknown references can be explained by a Renewal-theoretic model, which attributes to the cache lines a particular *generational behavior*. In simple terms, cache lines exist in "generations." A generation begins when a cache line is brought into the cache after a miss (see Figure 5.4). Immediately after the miss, a flurry of access activity ensues. While the cache line is being accessed, it is in its *live time*. At some point, the accesses cease and the cache line sits idle waiting to be evicted to make room for a new line. The time spent in this state is called the *dead time*. The fundamental characteristic of the generational behavior of cache lines, the reason why most of the cache is "empty" of useful

---

[6]Adaptive and compiler-assisted cache decay variants were also proposed in [127].

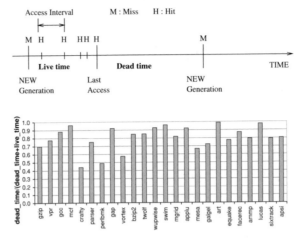

FIGURE 5.4: Generational behavior of cache lines (top). Dead time as a percentage of total time in the cache. Reproduced from [127]. Copyright 2001 IEEE.

data, and the reason why unknown references in trace sampling have a very high miss rate, is that (on average) *the dead time of cache lines by far exceeds their live time.*[7] Adapted from [127] the graph in Figure 5.4 shows that total dead time (for all cache lines) accounts for about 80% of the total time, on average, for the SPEC2000 benchmarks. This fundamental characteristic is exactly what cache decay tries to exploit.

Cache decay tries to guess whether a cache line is live at any particular point in time. It does this by measuring elapsed time since the last access to the cache line. Since the cache line is accessed only during its live time—which is typically short—it follows that the inter-access time between two consecutive accesses should be particularly short. It is easy then to guess when a cache line is *not* in its live time: if sufficient time has passed without an intervening access then most likely the cache line has entered its dead time and awaits eviction.

This approach is supported by data showing that the distribution of the inter-access time intervals, *measured in cycles*, is—for all practical purposes—*bimodal*. Figure 5.5 shows the distributions for three of the SPEC2000 benchmarks, gcc, compress, and vortex. The vast majority of the inter-access times are clustered around very small values (few hundreds of cycles or less), while a single cumulative bar at the far right represents the very large inter-access intervals. The middle parts of the graphs are practically empty. Similar inter-access time distributions are typical for many programs.

---

[7]So, for a number of unknown references, starting at a random point in the trace, it is unlikely to catch most of them in their live time. It follows that the majority will be cold misses!

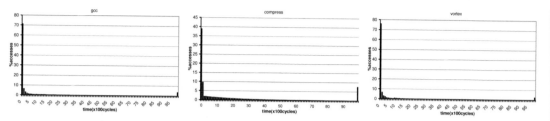

FIGURE 5.5: Inter-access interval distributions for gcc, compress and vortex from the SPEC2000 suite. Reproduced from [127]. Copyright 2001 IEEE.

The main task of cache decay is to gauge the idle time of a cache line in relation to its inter-access times. When the idle time of a cache line exceeds a limit called the *decay interval* (which is set to be beyond the cluster of the small inter-access times), the cache line is predicted to be in its dead time and is shut off using a gated-V_{dd} sleep transistor.[8]

Implementation: While there are a few possible implementations for cache decay (including some analog varieties), one of the better known methods uses a scheme of hierarchical counters [127]. The idea is to use counters in the cache lines to measure their idle time. A counter works like a stopwatch: it starts ticking after an access; if the cache line is accessed, it is reset; if, however, it ticks uninterrupted until it reaches the decay interval then the cache line is pronounced "dead." It is evident from Figure 5.5 that the idle time, the decay interval, needed to safely determine entry into the dead time is of the order of several thousand cycles. A counter maximum in the thousands would incur too much overhead, however, to include with each and every cache line.

The solution is to use much smaller, coarser-grain counters in the cache lines (Figure 5.6). These small counters advance every few hundreds (or even thousands) of cycles rather than every single. The beat is given by a single global cycle counter which counts these larger intervals. So, for example, if the global counter counts 1024 cycles and the local cache line counters are 2 bits, then they count decay intervals up to 4×1024 cycles. This scheme minimizes overhead, since the global counter can be easily piggybacked on cycle counters commonly found inside processor cores and the local cache line counters can be implemented asynchronously—possibly with an efficient coding such as Gray coding—to minimize switching overhead [127].

---

[8]It is interesting to note that cache decay works similarly to the way most other electronic devices are put into a sleep mode: by detecting that the device is idle for a period larger than its average interactivity time. Some typical examples are the hard disks and the laptop displays which are shut down by the operating system after preset periods of inactivity.

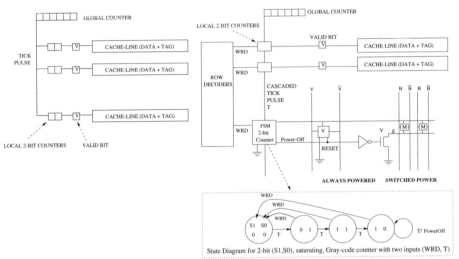

FIGURE 5.6: High-level view of the hierarchical counters. Reproduced from [127]. Copyright 2001 IEEE.

What makes this scheme viable, though, is that its effectiveness remains the same as if we had full cycle counters in every cache line.[9] In other words, it does not matter whether we count in cycles or in *thousands* of cycles when we are measuring large decay intervals.

The use of a global cycle counter also offers another advantage. It allows the flexibility to adjust the decay interval at a central, global, point. The local cache line counters measure idle time but they do not explicitly compare to the decay interval. Instead, when they reach their last state they consider the cache line to be dead and "activate" (turn *off*) the sleep transistor. Thus, the only way to adjust the decay interval is to control the period of the global beat (which ticks the local counters) given by the global counter.

Switching off power to a cache line has implications for the way the cache works. The first access to a powered-off cache line results in a miss (since data and tag are lost without power). The miss resets the cache line counter and restores power. Since the cache frame is written after the miss, the time to fetch the data from the lower part of the cache hierarchy is typically sufficient for the memory cells to stabilize after power is restored and accept the write.

☞ *decaying dirty data*: In a non-state-preserving technique, one has to address the problem of dirty data in a writeback cache. Decaying a dirty cache line necessitates a writeback. If the global counter signal is distributed in parallel to all the local counters, there is a chance that multiple lines will need to be written back at the same time if they all decay at the same global beat. This would choke the cache, in terms of access bandwidth requirements. To avoid this

---

[9]Even 2-bit local counters—very coarse time resolution—approximates the effect of full resolution.

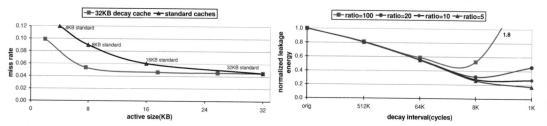

FIGURE 5.7: *Left*: Active size versus miss rate for decay caches. *Right*: Normalized leakage energy as a function of the decay interval for various *L2Access:leak* ratios. Reproduced from [127]. Copyright 2001 IEEE.

situation the global signal is distributed serially from one cache line to another, giving each line the chance to complete its writeback before proceeding to the next. Forcing an early writeback of dirty data is not necessarily bad for performance. In fact, prior to cache decay, Lee, Tyson, and Farrens proposed an "*eager*" writeback technique that yielded performance benefits by not bothering the cache with writebacks when it is servicing performance-critical misses [151].

Results: Cache decay has been extensively tested using the SPEC2000 in many cache configurations (e.g., instruction, data, L1, L2, direct mapped, set associative, and for many cache sizes). Overall, decay is very successful in switching off a significant part of the cache, on the order of 70% for Level-1 caches, impacting a minimal performance penalty of a few percent (less than 4%) [127].

Figure 5.7 shows a comparison of a single 32KB decay cache with standard caches of various sizes (4K, 8K, 16K and 32KB). What changes in the decay cache is the decay interval going from infinite (far right) to 1024 cycles (1Kc) at the far left. As the decay interval changes, the "*active size*" of the decay cache—the average part that remains powered on—and its miss rate are plotted. The decay curve is consistently below the curve for the standard caches. This graph shows that a decaying cache is always better than a standard cache: for the same size, the decay cache has a lower miss rate or, alternatively, for the same miss rate it has a smaller active footprint. This is a result of selectively keeping *active* the most important items in the decay cache.

However, decay can also result in energy and performance penalties. This is reflected in the graph on the right of Figure 5.7 which shows the overall benefit of decay in the form of *normalized leakage*. Normalized leakage is the ratio of the new leakage divided by the old leakage. The old leakage is the leakage before decay; the new leakage is the leakage with decay but augmented with the additional dynamic power consumed by the extra switching to implement decay *and* the extra power due to decay "mistakes." Because of the destructive nature of the gated-V_{dd} mechanism, mistakenly switching off a cache line results in an additional L2 access. These are called *decay-induced misses* and cost not only in energy (reducing the energy benefit of decay) but also in performance.

The more aggressive the choice of decay interval, the most mistakes are experienced. The larger the energy cost of a mistake (L2 access), the less the overall benefit from decay. To reflect this dependence, the right graph of Figure 5.7, plots several curves, each corresponding to a different relative cost for an L2 access. The ratio that specifies the relative cost is the L2access:leakage ratio, defined in [127]. As decay moves toward smaller decay intervals (and with more costly L2 accesses), it starts to lose its benefit and may even go into negative territory in the extremes. However, with a decay interval of 8000 cycles or more (for the simulated systems in [127]), decay-induced misses are so few that the relative cost of an L2 access becomes irrelevant. It is around these decay intervals where decay provides its maximal benefit.

Further results can be found in Hanson's work with Hrishikesh, Agarwal, Keckler, and Burger [93]. Hanson's work is one of the most detailed and extensive studies on cache decay and provides comparisons with two other leakage-saving techniques for caches. A detailed technical report by Hanson et al. greatly expands on the initial results reported for decay [94].

☞   *directions to improve cache decay*: Cache decay is based on the generational behavior of cache lines, and as such sits on a robust foundation. This allows it to work well in a wide range of conditions. There is room for improvement, though, over the initial proposal, on a number of aspects:

- *Decay-induced misses.* Cache decay has become synonymous with non-state-preserving techniques because of its use of the gated-V_{dd} mechanism to turn off cache lines. By mistakenly turning off cache lines in their live time, caches incur *decay-induced* misses that hurt both the energy savings as well as performance. In retrospect, decay can be thought of as the technique to detect dead lines in the cache. What to do with this information is a whole different matter: dead lines can be turned off (gated-V_{dd}), put into a drowsy mode, replaced, compressed, duplicated for reliability, etc. In a more recent work, the decay *policy* of detecting dead lines is used in conjunction with the drowsy *mechanism* that puts these lines into a low-leakage state.

- *Measuring time in cycles.* The decay interval is measured in cycles, which is a quantity that depends on architectural features. This makes it difficult to reason about decay intervals, especially across different programs, or in the same program but across different platforms. A better choice would be a more independent "time" metric such as the number of intervening accesses between two consecutive accesses to the same cache line. This is entirely a property of the application and does not depend on architectural parameters. The use of alternative counter-based time metrics is explored by Kharbutli and Solihin for managing replacements [135] but can be easily extended to manage leakage.

- *Need of the unfiltered reference stream to reliably detect generational behavior.* Decay works well only when the distribution of inter-access times is bimodal as is in the L1. Things get muddier in the L2 where the generational behavior of cache lines is obscured because of L1 filtering. In fact, what is observed in the L2 is the behavior of L1 conflicts or the generational behavior of lines which are accessed on a much different time scale than the L1 lines. Despite the initial assessment that decay works well in the L2, albeit with very large decay intervals, Abella et al. exposed the problems and proposed a new approach for L2 decay [1].

5.2.3   Adaptive Cache Decay and Adaptive Mode Control

Although cache decay is capable of shutting off a significant part of the cache with a small performance impact, an aggressively small decay interval can cause a jump in the number of decay-induced misses, destroying its advantage over cache resizing. On the other hand, a conservatively large decay interval misses the opportunity to turn off cache lines already in their dead time. Cache decay also carries a fixed overhead over an oracle prediction for dead lines. This is because cache decay has to *wait* for the length of the decay interval from the last access to a (dead) cache line to shut it off. This "missed opportunity" to save leakage increases with larger decay intervals. In contrast, an oracle prediction knows immediately when a line enters its dead time and wastes no time to start saving leakage.

It is clear that tuning the decay interval is critical in making decay work well for different applications, or even for different phases of an application. Zhou et al. found, by trial and error, that decay intervals vary significantly for SPEC2000 benchmarks. For instance, to keep the performance penalty below 4%, decay intervals in a 64KB four-way set-associative cache, range from 14 000 cycles for JPEG to 98 000 cycles for LI [250].

The selection of a decay interval for an application is thus a non-trivial task that must balance dynamic power increase and performance loss to gains in leakage savings. Furthermore, a single decay interval for an application derived from a profiling run is possibly not optimal for every input data set or even for different phases of the application. To avoid passing this burden to the user (programmers, compilers, operating systems), adaptive hardware mechanisms have been proposed to adjust the decay interval dynamically. While there are a number of proposals in the literature, here we describe three initial proposals: (1) local per-line decay interval adaptation [127], (2) global adaptation of the decay interval based on application performance feedback [250], and (3) a generalization of the global adaptation approach using control theory [219].

Per-cache-line adaptive decay: The idea behind this technique is to dynamically adjust the decay interval individually per cache frame. The criterion for adjusting it is based on the accuracy versus the aggressiveness of the dead line prediction. If the decay interval is too small (too aggressive), then a cache line might be predicted dead while it is still in its live time which leads to decay-induced misses. In this case, the decay interval is *increased*. In contrast, decay intervals that are too large—although very accurate in predicting dead lines—waste considerable time before saving leakage thus incurring a significant fixed overhead. In this case the decay interval is *decreased*.

There are two issues in implementing this technique: first an efficient way of increasing and decreasing the decay interval, *individually per cache line*, must be available. This is solved by halving or doubling the decay interval as this is the easiest thing to do (see "Sidebar: per-cache-line adaptive implementation details" for details). Second, a scheme for deciding whether the decay interval is too large or too small must be devised. This is accomplished by detecting when decay makes mistakes, that is, when it causes decay-induced misses. If a cache line is turned off and, as a result, a decay-induced miss ensues, the decay interval is too short and is increased. Going the opposite direction is opportunistic: in the absence of any decay-induced misses (dead lines are replaced by different lines), the decay interval is reduced in hope that this will decrease the fixed cost without incurring new decay-induced misses (see "Sidebar: Per-cache-line Adaptive Implementation Details" for further details).

☞   *per-cache-line adaptive implementation details*: There are two ways to change the decay interval independently for each cache line. The first is to change the number of active *counting* bits in the local counter of a cache line. The second, and the one adopted in [127], provide not one but several global signals, each "beating" at a different rate (e.g., double the previous one). This is as easy to do as to assign transitions of consecutive high-order bits of the global counter to different global signals. To change the decay interval locally in a cache line, the local counter simply *chooses* which global signal to feed from.

To change the decay interval, decay-induced misses must be readily detectable. This is straightforward if the tags remain powered up at all times. Hits to decayed lines are then evident. But it is also possible to empirically detect mistakes without having the tags powered-up [127]. The method proposed is to detect mistakes by gauging how soon after decaying a cache line its frame is accessed. This works for direct-mapped caches but still requires at least a few powered-up tag bits to disambiguate between different cache frames in set-associative caches.

The actual scheme proposed is shown in Figure 5.8. The local counter is double-booked to count time even *after* a line is decayed. The deciding factor is *the time when the next access arrives*. An access very soon after decaying a line, is taken to be a decay-induced miss (even if it is unclear if it is truly an access to the *same* cache line). The decay interval

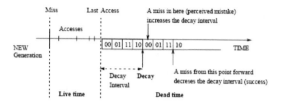

FIGURE 5.8: Adaptive per-cache-line decay. The decay interval is counted by the 2-bit local counter and the line is decayed when the counter reaches state 10 (note the Gray coding for switching efficiency). After the line is decayed the counter starts counting again but this time to gauge the arrival time of the next access. In this particular example, if the next access comes when the counter is in its first state (00) the (unknown) access is taken to be a decayed-induced miss and the decay interval is doubled; the decay interval does not change in the middle two states of the counter, while if the next access comes when the counter has saturated at state 10, the access is taken to be a replacement and the decay interval is halved. Reproduced from [127]. Copyright 2001 IEEE.

is *doubled* in this case. In contrast, an access that comes much later (after the local counter advances a few more steps) is taken to be an access to a different cache line—a replacement. In [127] if the miss occurs with the local counter stuck at its highest value, the decay interval is to be too large and is halved to reduce the fixed overhead; otherwise, it is left unchanged.

Global decay interval adaptive techniques: The concept of cache decay was independently invented by Zhou, Toburen, Rotenberg, and Conte, who proposed an adaptive policy to dynamically adjust the *global* (for the entire cache) decay interval [250]. Their adaptive scheme is called Adaptive Mode Control (AMC). Its adaptation is based on a performance feedback loop: the decay interval is adjusted so that the number of decay-induced misses—i.e., the performance loss—is bounded. In contrast to the per-cache-line adaptive decay which can only be controlled indirectly, this technique allows excellent control over the trade-off between performance and leakage-reduction.

The underlying mechanism to turn off dead cache lines is similar to cache decay (see Figure 5.9).[10] A two-tier hierarchy of counters, consisting of local counters per cache line (called *Line Idle Counters*, or *LICs*) and a global counter (called *LIC update interval counter*) measure the decay interval. In contrast to cache decay, the global counter (the LIC update interval counter) remains fixed. The global LIC update interval is set to a value of 2048 cycles that is not too long to hamper decay, neither not too short to force frequent switching upon the local LICs. As each LIC advances, it is compared to a threshold. If the LIC exceeds the threshold the line is decayed.[11] The comparisons take place in each cache line using a local comparator called Mode

---

[10]AMC also uses a gated-V_{dd} mechanism to turn off cache lines which makes it a non-state-preserving technique but—like cache decay—this is a *choice*, not an inherent property of the technique.

[11]Recall that in cache decay the local counters do not compare against anything but simply decay the cache line when they reach their "last" (saturation) state.

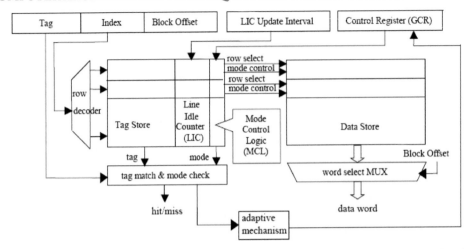

FIGURE 5.9: Adaptive Mode Control (adapted from [250]). The decay interval changes by changing the global control register (GCR) which holds the threshold for the local counters (LICs). The global counter (LIC Update Interval) does not change. Adaptation is controlled performance feedback loop (adaptive mechanism). Reproduced from [250]. Copyright 2001 IEEE.

Control Logic (MCL)—shown next to the cache lines in Figure 5.9. The threshold, which is the same for all cache lines, is kept in a global control register, the GCR.[12]

Adaptation of the decay interval is simply a matter of setting the GCR to the appropriate threshold for the LICs. For this, a feedback loop monitoring decay induced misses is employed. The feedback loop works by adjusting the decay interval to keep the decay-induced miss rate around a desired *target* which is given as a percentage of the application's *true* miss rate.

Decay-induced misses are distinguished from normal misses by keeping the tags powered on at all times and detecting accesses to decayed lines (i.e., accesses that would be hits if the line was still powered on). The decay-induced miss rate is measured within a time window, and at the end of the window the decay interval is adjusted. If the decay-induced miss rate is less than the target, the decay interval is decreased; otherwise, it is increased. Hitting the exact target is difficult to do. Most often, the feedback loop simply oscillates around the target, constantly adjusting the decay interval. To avoid this needless exercise, the decay interval is not allowed to change if the miss rate falls within a band enveloping the *target* miss rate.

Figure 5.10 shows the behavior of the feedback loop. Two bold lines in the graph represent, respectively, the "*ideal*" or *true* miss rate, which is the miss rate without decay, and the "*target error*" which is the *additional* miss rate of the decay-induced misses. The target error is defined as the fraction of the ideal miss rate via the *performance factor (PF)* which is externally

---

[12]Note that although the hardware can accommodate individual decay intervals per cache line, adaptation is based on a single global decay interval.

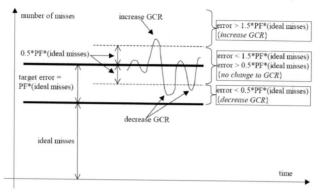

FIGURE 5.10: Feedback control for AMC. The performance factor PF defines a target error (decay-induced miss rate) as a percentage of the "ideal" miss rate (miss rate without decay). The actual miss rate is allowed to move inside a band around the target miss rate (ideal misses + target error) without changing the decay interval. But when the actual miss rate ventures outside this band the decay interval is changed (increased if the actual rate overshoots the upper band limit or decreased in the opposite case) to bring it back into the limits. Reproduced from [250]. Copyright 2001 IEEE.

supplied. The decay interval does not change as long as the miss rate of the cache remains within a specific range (band) around the target error. If the miss rate ventures outside the band limits, the decay interval is changed to bring the miss rate back inside.

Zhou et al. report that AMC yields similar benefits to adaptive cache decay but with a much more robust control over performance. Whereas adaptive per-cache-line decay settles in a power-performance point that is controlled only indirectly, AMC sets a specific ceiling in performance loss and adjusts decay under that constraint.

Control theoretic techniques: Velusamy, Sankaranarayanan, Parikh, Abdelzaher, and Skadron formalized the Adaptive Mode Control technique using control theory [219]. The contribution of their work is to develop a formal method for feedback-control for cache decay. The method is also based on identifying and controlling decay-induced misses by monitoring the tags.

Like ACM, the tags are kept powered-on at all times. Velusamy et al. show that both their formal method and AMC adapt very well to changing application behavior as opposed to statically selecting a fixed decay interval. The latter leads to unwelcome behavior during different application phases. The formal controller, called Integral Miss Control (IMC), is almost as good in practice as AMC but easier to design and tune. This gives a strong incentive to use control-theoretic approaches when designing closed-loop responses.

Comparison: To summarize, Table 5.2 shows the main differences between the adaptive decay methods. Although the per-cache-line adaptive decay is finer-grain and thus more

TABLE 5.2: Comparison of Per-cache-line Adaptive Decay and AMC

	Per-cache-line Adaptive	AMC & Control Theoretic Methods
Adaptation granularity	Individual decay interval per cache frame	Single decay interval for the whole cache
Adaptation method	Based on balancing two opposing trends: 1. Minimizing decay induced misses by increasing the decay interval in the cache frames where they are detected. 2. Minimizing the fixed overhead by decreasing the decay interval in frames where there is an absence of decay-induced misses.	Performance feedback loop: Adjust decay interval to bring decay-induced misses within a region around a specific target
Control	Fuzzy: Balancing the above two opposing trends is done by judging how soon or how late the next miss to a decayed cache frame occurs.	Specific: the performance feedback loop can be accurately controlled by setting exact targets and limits
Tags	No need for powered-up tags to detect decay-induced misses which can be done with a time-proximity heuristic. However, for full accuracy tags need to be powered-up. Exception: A few tag bits are needed to disambiguate among accesses to cache frames in set-associative caches.	Need for powered-up tags to detect decay-induced misses.
Overhead	Multiple global interval signals must be available for the local counters to choose from. Additional control logic is required per cache line to judge decay-induced misses and adjust the local decay interval accordingly by choosing a global signal.	Decay threshold comparators must be available per cache line. Performance feedback loop needs its own block of logic.

appealing as an adaptive technique, in reality it is harder to control than AMC. Besides, there is nothing in the AMC (or its control-theory extension) to preclude finer, per-cache-line, adaptivity although something like this has not been evaluated as yet. Conversely, control in the per-cache-line adaptive scheme could be improved by establishing a feedback loop similar to that of AMC.

5.2.4   Decay in the L2

The generational behavior of cache lines, upon which cache decay is based, is readily apparent in the L1. Unfortunately, this is not so in the L2, where it is obscured by the L1 filtering of the accesses. In reality, the L2 observes the behavior of the L1 *misses*.

The typical generational behavior for a cache line is to experience a flurry of accesses—with short inter-access times—when it is first brought into the cache, followed by a long dead time until it is replaced. The situation in the L2 is different. In the absence of a conflict (in the L1), only a single access to a cache line, at a time, is visible in the L2: the one that transfers the cache line to the L1. No more accesses are visible in the L2 until the cache line is replaced in the L1. The next L2 access corresponds to a *new* L1 generation. In the case of a conflict, the L2 observes accesses at random points in a cache line's generation as a result of two different generations elbowing each other out of the L1 while still in their live times.

The end result is that the decay intervals in the L2 must be very large for the technique to work reliably without significant performance degradation. The initial studies concluded that cache decay works well in the L2 using such large-decay intervals (of the order of 1M cycles). However, this could be attributed more to the easy fit of the benchmarks in large L2s (leaving ample empty space which can be turned off) rather than to the accuracy of the technique in determining dead lines. With a scaled-down L2, decay is stressed to provide a good turn-off ratio for a reasonable performance loss.

A better adaptive decay for L2: Abella, González, Vera, and O'Boyle exposed the problems in the L2 and proposed a new technique to turn off unused L2 lines [1]. The goal is to find an appropriate method to dynamically adjust the decay interval for the L2. Abella et al. make a key observation: the inter-access time—the time between two accesses to the same cache line—in the L2 seems to depend on the *number* of accesses to the cache line. Thus, the decay interval for each cache line needs to change as the cache line is accessed. The actual behavior of the inter-access times, as a function of the number of accesses to a cache line, depends on the application and needs to be captured at run-time. The implementation details can be found in [1].

Although this technique carries significant book-keeping overhead, its benefits more than make up for it. The decay interval as a function of the number of accesses to a cache line is much more effective than the very large static decay intervals typically needed in the L2 to avoid

TABLE 5.3: Non-inclusion Policies. Adapted from [154].

Policy	Sub-block is Decayed	Sub-block is Reactivated
Conservative	When L1 block becomes DIRTY	On writeback
Speculative-II	When block is transferred to L1	When block is accessed in the L2 (even for writeback)
Speculative-IV	When block is transferred to L1	When L1 block is evicted

severe performance loss. Because this decay adaptation takes into account how the inter-access times vary with the number of accesses, it also outperforms both the adaptive per-cache-line decay and the AMC technique.

A matter of inclusion: Li, Kadayif, Tsai, Vijaykrishnan, Kandemir, Irwin, and Sivasubramaniam examine L2 decay from a different perspective [154]. They make the observation that cache lines do not have to be powered-on in the L2 while they are live in the L1. In essence, they make the cache hierarchy non-inclusive for live lines. Their proposed policies work at the granularity of L1 lines. In case the L2 lines are larger, they are divided into sub-blocks equal to L1 lines; but to simplify the discussion here, we assume that the L1 and the L2 line sizes are the same. Gated-V_{dd}, as usual, turns off individual L2 lines.

Li et al. propose several policies and also compare the use of decay (non-state-preserving) and drowsy (state-preserving) techniques. The latter are discussed in Section 5.3. The proposed policies differ on when an L2 line is decayed and when it is reactivated. Table 5.3 lists the three non-state-preserving policies using the terminology of the Li et al. paper [154], while Figure 5.11 shows graphically their behavior.

The conservative policy only decays L2 lines that are no longer needed because they are modified in the L1 (Figure 5.11). On a writeback, the L1 line is put back into the L2. The other two policies (Speculative-II and Speculative-IV) decay an L2 line as soon as it is transferred to the L1. In Speculative-II, an evicted L1 line does not go back to the L2 unless it is dirty. If it is dirty, the L2 line is reactivated to receive the writeback. Otherwise, the L2 line is powered-up *on demand* upon the next access which exposes the full penalty of an L2 decay-induced miss (Figure 5.11). In contrast, in Speculative-IV the L2 line is reactivated when the L1 line is evicted (even when it is clean). When the L2 is accessed again, it is likely that the line is ready and waiting to be transferred back to the L1, hiding the reactivation cost.

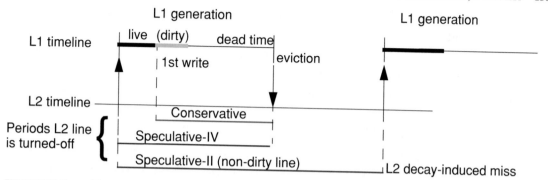

FIGURE 5.11: Time diagrams for the non-inclusion policies. Adapted from [154].

Not surprisingly, simulations show that the conservative policy does not save as much leakage as the others but, on the other hand, does not hurt performance either. Speculative-II saves the most energy by being very aggressive, but degrades performance significantly hurting the resulting EDP. Although Speculative-IV saves less leakage due to early reactivation, it also avoids going to the main memory as extensively as Speculative-II does. Thus, it consumes less dynamic energy, performs better and yields a better EDP than Speculative-II. It is also possible to combine cache decay in the L1 with an L2 policy (especially if it is state preserving) for significantly better results than just having cache decay at both levels [154].

5.2.5   Four-Transistor Memory Cell Decay

Decay in static RAM mimics the fleeting nature of dynamic RAM: data that are not touched or refreshed, are lost. The idea of using four-transistor quasi-static cells for decay stems from this parallel. Hu et al. make the observation that much of on-chip storage is devoted to transient, often very short-lived, data. Despite this, virtually all chip designs use array structures based on six-transistor SRAM cells that store data indefinitely (Figure 5.12, left side). A charge stored

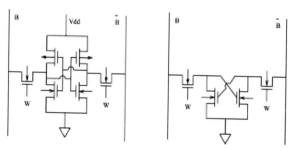

FIGURE 5.12: 6T versus 4T memory cell. Adapted from [106].

in a 6T cell will be maintained as long as that cell is connected to the source and drain voltages (V_{dd} and ground).

But leakage-control techniques, such as those discussed in Section 5.3, carry an overhead (per entry) that cannot be easily amortized in smaller structures or structures with very small entries. Instead, Hu et al. entertain the idea of using quasi-static, *self-decaying*, four-transistor (4T) memory cells, especially for branch predictors and branch target buffers (BTBs) [106].

The reasoning behind this approach is that quasi-static 4T cells (Figure 5.12, right side)—which have no connection to V_{dd}—provide decay functionality inherently: if not accessed, their charge gradually leaks away at a rate that is a function of the cell's specific design and operating temperature. Conversely, they are recharged upon access. 4T cells are, therefore, uniquely well-suited for decaying structures, such as branch predictors and BTBs, where the overhead of a 6T-cell decay technique would be prohibitive.

BTB and predictor structures—This behavior is conducive for decay, since the data held is both transient and predictive. It is transient in the sense that data which has not been used for a sufficiently long time has decayed. It is predictive in the sense that allowing a value to leak away even if it will be used again does not harm correctness. Using a decayed value *may* merely cause a misprediction that can be handled by existing hardware. This is a key difference from caches, where using decayed data leads to incorrect execution of the program.

The key design points for implementing decay in branch predictors and BTB's are the decay interval—which can be affected by circuit design—and the granularity of access—how many entries are accessed simultaneously because of the layout of the branch predictor or BTB structure. Their combination determines the active ratio and thus, the leakage savings and the resultant performance [106].

4T decay is met with some skepticism because it is an analog design with many factors that could affect it. Many issues, including metastability issues, dependence on temperature, etc., need to be addressed to deploy such a technique in practice. It does hold the promise, however, for very low leakage since only the memory cells that are actively used can dissipate leakage power.

5.2.6   Gated V_{dd} Approaches for Functional Units

The first leakage-control policies, using gated V_{dd} (e.g., DRI, cache decay, AMC), target caches. In this context, the main energy trade-off is between the saved leakage energy and the extra dynamic power consumed by misses on the turned-off cache lines. The extra energy cost of switching the sleep transistor on and off was not of immediate concern. This, however, was of little consequence, since the timescales for inactivity (and therefore the potential leakage savings) are such that they could easily absorb the sleep transistor switching cost without difficulty. In other words, the cost of switching the sleep transistor and recharging the internal nodes of the memory cells when restoring power is small in comparison to the hundreds of thousands of cycles of leakage energy saved for each cache line. A small change in the decay interval—maybe a few tens or a few hundreds of cycles—can more than make up for this cost.

Things are different, however, when it comes to power-gating functional units. There, the time scales are entirely different. Inactivity must be exploited in a matter of 10's of cycles (not 100's of *thousands*). In addition, there is no *substantial direct* energy cost, such as the decay-induced miss in the caches, when making a mistake and switching off a needed functional unit.[13] Instead, what matters the most in functional units is the dynamic energy cost in powering them up or down.

Hu, Buyuktosunoglu, Srinivasan, Zyuban, Jacobson, Bose [105] provide an excellent analysis of the costs involved when power-gating functional units. They assume that a functional unit is power-gated either with a header (connected to V_{dd}) or a footer transistor (connected to ground). In their analysis, they include both the cost of switching the gating transistor and the cost of recharging a functional unit in relation to the length of time it was discharging. Instead of calculating the end result for a specific example, making a number of assumptions for the parameters involved, they give analytic formulas that yield the break-even point, in cycles, for power-gating a functional unit. To simplify the formulas, a *leakage factor L* is introduced, which specifies the ratio of the average leakage power to the average switching power dissipated per cycle by a functional unit.

Having an accurate framework for assessing power-gating at a fine grain, Hu et al. examine two policies for power gating functional units. The first is a time-based policy, inspired by cache decay, while the second is an event-guided policy and in particular a branch-prediction guided policy.

Functional unit decay: The first approach works similarly to cache decay. If an idle period is detected in a functional unit then it is switched-off. The are three timing factors that determine the behavior of this approach: the break-even point in cycles after which there are net gains in energy; the time it takes for the functional unit to wake up from the moment it is needed; and the decay interval, i.e., the time it takes to decide to put the functional unit in sleep mode. The first two are technology and functional-unit specific, while the third, the decay interval, is an architectural knob that one can turn to tune the policy.

- The first timing factor, the break-even point, varies depending on the technology and functional unit. In particular, it varies depending on the leakage factor L, the ratio of leakage energy to dynamic energy.

---

[13]But there can be indirect energy costs due to power-gating mistakes which reduce performance.

- The second factor, the time it takes to wake up a powered-down functional unit, depends on the capacitance that needs to be recharged when power is restored. The wake up factor determines the performance impact of the technique. The more often a functional unit is *not ready* when it is needed, the worse the performance impact. Thus, to minimize the negative performance impact, one has to power-down a functional unit the least number of times possible. This is the most pessimistic scenario that assumes that there is no advance warning that a functional unit might be needed. If an advance warning were available, the wake up time could be hidden resulting in little or no performance penalty.

- Finally, the decay interval determines the idle periods that are "selected" for powering down the functional unit. A very small decay interval powers down a functional unit even in very short idle periods; but this can be more damaging than beneficial depending on the other two timing factors. In contrast, a large decay interval filters out the short idle periods and powers down the functional unit only in larger idle periods.

Different types of functional units differ on the first two factors but also in the behavior of their idle periods. In general, the larger the capacitance of a functional unit, the larger the break-even point and the wake up time. Floating point functional units tend to have a wide range of idle periods (in SPEC-FP 2000). Although their short idle periods are more numerous than their longer ones, most idle cycles are due to the longer periods by virtue of their size. In this case, a long decay interval skips the short idle periods, and selects only the large ones. This minimizes the number of times the functional units are unavailable because they are powered down while still benefiting by having the functional unit powered down for a significant part of the time. Overall, this technique can power-down the floating point units for 28% of the time with only minimal performance penalty (less than 2%) for the SPEC FP 2000.

Things are much harder with the integer functional units. These are used much more often—for instance, loads that need address arithmetic are issued every four or five instructions. This means that the bulk of the idle periods are short. Longer idle periods are rare and basically appear after L2 misses where the processor may be starved of integer operations. The end result is that the integer functional units can only be powered-down for a few percent of their time (6–8%) without significantly harming the performance. Too small a decay interval increases the power-down time but hurts performance, whereas a longer decay interval fails to capture the numerous small idle periods, but, on the other hand, does not allow too many chances to catch the functional units in their sleep.

Event guided power-gating. To improve power-gating, especially for the integer functional units, Hu et al. propose using various events as clues to upcoming idleness of the functional units. For example, upsetting events such as L2 misses, instruction cache misses, or branch mispredictions all point to idleness of the functional units until the processor pipelines recover and return to full operation.

Hu et al. concentrate on branch mispredictions. Upon detecting a mispredict, the functional units are put immediately into sleep mode without waiting for the normal decay interval. This simple rule extends the powered-down time of the functional units without incurring any additional performance penalty. The use of clues increases the percentage of cycles in sleep mode for a given performance loss, or, conversely, for the same percentage of cycles in sleep mode the use of clues eases the performance impact. Similarly to branch mispredictions, other events can also provide useful hints for the idleness of the functional units but have not been studied further.

5.3   ARCHITECTURAL TECHNIQUES USING THE DROWSY EFFECT

One disadvantage of the gated V_{dd} mechanism is that it destroys state. The first approaches to control leakage based on this mechanism (DRI cache, cache decay, AMC, etc.) are known as non-state-preserving. In the case of cache decay and related approaches, the reasoning is that most of the lost state is useless anyway. And that would be fine if it were not for the problem of mistakes—decay-induced misses—which actually harm performance. The drowsy effect was proposed to address this problem, introducing a new class of *state-preserving* leakage-reduction techniques.

5.3.1   Drowsy Data Caches

In response to the gated-V_{dd} problem of losing state, Flautner, Kim, Martin, Blaauw, and Mudge proposed another approach to curb leakage in memory cells [77]. The drowsy mode is a low supply voltage mode for the memory cells, i.e., Dynamic Voltage Scaling (DVS) for leakage.

Similar to the DVS approaches discussed in Chapter 2, this type of DVS also has to do with idleness; but not with frequency scaling. Memory cells which are idle, i.e., are *not* actively accessed, can be voltage-scaled into a drowsy mode. In this mode, transistors leak much less than with a full V_{dd} as explained in Section 5.1.1. Figure 5.13 shows the design of a drowsy cache from [77]. A "drowsy" bit controls the two levels of supply voltage (V_{dd} or V_{ddLow}) to the memory cells of a cache line. Memory cells are in drowsy mode when fed from V_{ddLow}.

The leakage reduction of the drowsy mode is not as profound as that of the gated-V_{dd} approach which completely cuts off the path to V_{dd} (or, equivalently, to ground). However, allowing for some nonzero supply voltage preserves the state of the memory cell. This happens as long as the supply voltage is strong enough to replenish the charge in the cell's internal nodes. However, a memory cell in drowsy mode cannot be accessed with the full-V_{dd} circuitry of the

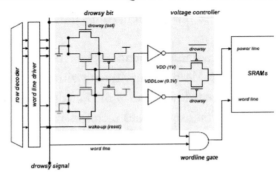

FIGURE 5.13: Drowsy cache. Reproduced from [77]. Copyright 2002 IEEE.

cache. It first has to be voltage-scaled back to full V_{dd}. Because this is not instantaneous, there is a penalty, albeit small, in accessing drowsy cells.

High-level policies for drowsy caches: Because state is preserved in drowsy mode, there is no danger in experiencing long miss latencies when accessing drowsy cache lines. The penalty to voltage-scale a drowsy cache line back to full V_{dd} is relatively small—a few (single-digit) cycles. Whereas it would matter significantly which cache lines are put into low-leakage mode in a non-state-reserving technique, with the drowsy mode it does not matter; mistakes cost very little. This makes sophisticated techniques that determine the idleness of cache lines unnecessary, especially if one factors in their dynamic power cost. Flautner et al. thus propose a very simple policy—fittingly called *Simple*—for the drowsy mode: the whole cache is periodically put into drowsy mode—all of the cache lines regardless of usefulness or idleness. The small percentage of active cache lines are going to exit the drowsy mode, on demand, incurring a small latency penalty. Since this latency is experienced on hits, programs which are sensitive to hit latency are going to be hurt the most. A variable hit latency can also complicate instruction scheduling in an out-of-order core, further degrading the performance [180].

The simple policy is quite effective: it can put into drowsy mode 80–90% of a 32KB L1 data cache while incurring a slight performance penalty of 1%. These numbers are for a four-instruction wide out-of-order core and assume a very aggressive one-cycle penalty for accessing drowsy cache lines. The Simple policy does not perform as well with instruction caches which need to be handled differently.

Improvements on the drowsy policy: Petit, Sahuquillo, Such, and Kaeli [181] improved on the Simple policy of Flautner et al. by applying few smart heuristics. Their approach is to maintain the low complexity of the initial idea by adding very little hardware.

The goal is to improve on the Simple policy which blindly puts all cache lines in drowsy mode. In the Simple policy no effort is spent to distinguish between active (important) and idle (useless) cache lines. On the other hand, precisely determining the individual status of each and every cache line, la cache decay, veers off the desired course of simplicity. Instead, Petit et al. propose simpler heuristics to *filter* the lines that are put in the drowsy mode.

Considering the cache lines in an associative set, it is obvious that if there is a *live* cache line among them it must be the most recently used (MRU) line. Thus, the first policy is to exclude the MRU line from going into drowsy mode. In fact, the policy allows *only* the MRU line to remain awake in the set. The policy is called MRU ON, or simply MRO. Upon a hit on a drowsy line, the line is woken up and becomes the MRU line. The previous MRU line is put in the drowsy mode. According to Petit et al., 92% of the cache hits are hits on the MRU line so a change in the MRU status, which incurs dynamic switching, is rare.

The second policy simply keeps awake the two most recently used lines per set inasmuch as both are good candidates for being active (alive) rather than idle (dead). The policy is called Two-MRO (TMRO). This choice is justified because most of the remaining 8% of the hits that do not hit the MRU line are accesses to the second most recently used line. Since more lines per set are kept awake energy savings are less than MRO but accesses to drowsy lines are minimized.

Both these policies try to minimize the performance penalty from accessing drowsy lines. This penalty, however, is only important if the transition from drowsy to active takes a few cycles. Petit et al. examine these two policies with a transition latency of three cycles which is reasonable (if not overly aggressive) for a number of designs.[14]

Petit et al. conduct experiments which show these two simple policies, MRO and TMRO, offer power savings of about 72% and 48%, with a hit ratio on active lines of 92% and 98%, respectively. However, neither can adapt to the dynamic access behavior of various applications. For this reason, the authors propose an adaptive variation that dynamically selects, per set, one of the two policies depending on the number of accessed lines in the set within a time window. This adaptive policy called Reused-MRO (RMRO), settles midway between the MRO and TMRO and outperforms the Simple policy of Flautner et al. both in leakage savings (65% versus 62%) and in the hit ratio on active lines ($\sim$99%).

5.3.2 Drowsy Instruction Caches

One situation where the Flautner et al. Simple policy—periodically putting the whole cache in drowsy mode—does not work very well is the instruction cache [138]. This is because instructions exhibit strong temporal and spacial locality and any delay in fetch shows up immediately on performance. Putting active instructions in drowsy mode is a sure way to lower

---

[14]The wake-up latency depends on how big is the cache line (number of cells, length of power lines) but more importantly how large is the voltage controller. To drop the latency to the overly aggressive 1 or 2 cycles in the initial proposal [77], a quite large controller (128 to 256 L_{min} wide) must be used, significantly increasing the overhead per cache line [137].

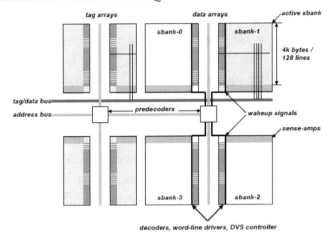

FIGURE 5.14: Banked cache organization. Only actively accessed banks expend dynamic power and are in full V_{dd}; the rest of the banks are put in drowsy mode. Reproduced from [137]. Copyright 2004 IEEE.

performance. One solution is to apply a decay policy, but using the drowsy mode instead of gated-V_{dd}. This way, active instructions would remain at full V_{dd} and incur no additional latency.

In a more recent paper, Kim, Flautner, Blaauw, and Mudge return to the Simple policy for the instruction caches with a renewed study on the locality characteristics of instructions [139]. Their results show that the Simple policy can also work on instruction caches albeit at a much different window than for a data cache. In the same paper, they also propose a novel version of the drowsy circuit requiring a single V_{dd} power supply [139]. Here, we discuss two other approaches specifically tailored to take advantage of the code behavior.

Leakage control at the bank level and next-bank prediction: Kim et al. examine a different approach which is a variation of the Simple policy but tailored to instruction caches. First, the granularity of the drowsy mode is changed from the cache-line level to the cache-bank level. Kim et at. point out that banking is fairly common in many cache implementations. It is often used for reducing dynamic power consumption by pre-decoding addresses and enabling only the relevant bank for an access (see Section 4.9 and the work of Ghose and Gamble [83]). Similarly, only the bank that is actively accessed is kept in full V_{dd}, while all other banks are put in drowsy mode (Figure 5.14). This works well because of high spatial locality in code: instructions are fetched in program order, conditional jumps tend to jump close by, and loops repeat the same code multiple times. An active bank tends to remain active for a long time.

The problems appear when functions which are resident in other banks are called or when active code spans more than one bank, or in general when there are frequent bank transitions. The solution in this case is to *predict* bank transitions with a technique called Next Sub-Bank Prediction (NSBP). Since bank transitions are due to branches—or the execution spilling over to the next bank—they can be predicted as well. Interestingly, Kim et al. discover that bank transitions are mostly due to function calls/returns and long distance jumps, which are, of course, highly predictable. It is a straightforward matter of using a CAM buffer or the cache tags themselves to keep the correspondence of such points in the code to bank transitions. Upon encountering a point that indicates a transition to another bank, the target bank is reactivated.

Prediction accuracy for a CAM buffer next sub-bank predictor ranges from 51% for 32 entries to 78% for 256 entries. The resulting performance penalty is small (less than 2% for the 32-entry buffer and less than 1% for the 256-entry). Things get better if the next bank prediction is associated with the tags. It is less costly and better performing than a 128-entry CAM buffer. It is also possible to do static next sub-bank prediction at compile-time or link-time completely eliminating the overhead of dynamic prediction [12]. The end result is that the NSBP drowsy policies in the instruction cache work equally well but with slightly less performance impact than state-destroying, gated-V_{dd} decay with a fixed decay interval.

Program hotspots and code sequentiality: Similarly to the ideas of Kim et al. at the bank level, Hu, Nadgir, Vijaykrishnan, Irwin, and Kandemir exploit code behavior but at a finer granularity [104]. Their approach is based on identifying the instructions comprising the working set of executing code. Such instructions are kept active, out of reach of leakage-control policies, until execution moves to a different working set.

The working set in this case corresponds to a program phase. Typically, program execution occurs in phases. A program phase is identified by instructions which exhibit high-temporal locality in the duration of the phase. In general, not all such instructions are spacially close but can be scattered across the address space. If a program phase persists long enough, it is considered to be a *hotspot* [104].

Whereas the approach of Kim et al. is at the bank level, assuming that a loop body maps on a cache bank and occasionally makes calls to subroutines mapped on other banks, the approach of Hu et al. is at a much finer grain: cache lines containing the hotspot instructions are *individually* marked as such regardless of where they are in the cache. These lines are then excluded from leakage control.

Marking the hotspot instructions relies on an application's branch behavior. In particular, it is accomplished using information from the Branch Target Buffer (BTB). The BTB identifies the basic blocks that comprise a hotspot by keeping statistics on how often the basic blocks are executed. For each BTB entry two basic blocks are traced: the basic block that starts at the target address and the basic block at the fall through address (when the branch is not taken). Statistics are kept in frequency counters associated with each BTB entry and are collected during a time window. When a frequency counter exceeds some empirically chosen threshold, the corresponding basic block is considered hot. All ensuing fetches up until the next BTB access are tagged as hotspot cache lines.

The leakage-reduction policy is the Flautner et al. Simple policy. At the end of a time window, all the cache lines are put into drowsy mode *except the lines that are tagged as hotspot lines.* The protection of these lines from the leakage-control policy is immediately revoked and has to be re-established in the current window anew. In addition, BTB frequency counters are halved (by a 1-bit shift) with the end of each window to allow the "hotspot" working set to gradually change. A new time window can start sooner than its preset time interval if a loop is detected. In this case, there is no need to wait till the end of a full time interval to detect additional hotspot instructions.

Similar to the next-bank prediction of Kim et al., which tries to hide the re-activation latency, Hu et al. also propose—at a much finer granularity—*just-in-time activation* of individual cache lines. Since their proposal works at the cache line level, a simple sequential activation mechanism, that activates the succeeding cache line (succeeding index) from the one that is accessed, takes care of the straight-line code. However, way prediction is needed in set-associative caches to avoid waking up a whole set [104].

The most sophisticated scheme proposed by Hu et al. employing hot-spot detection, just in time cache-line activation, and bank activation to detect spatial changes in the working set, outperforms the coarse-grain technique of Kim et al. at the bank level, as well as a compiler approach discussed in Section 5.3.6. This is not surprising since at the cache line level there is potential for much better energy savings while maintaining the full performance advantage. The proposed scheme results in a 63% reduction in EDP over the unoptimized base case, 48% reduction over the bank-level technique, and 38% over the compiler-level technique.

5.3.3 State Preserving versus No-state Preserving

The proposal for the drowsy cache was put forth to address the main weakness of the gated-V_{dd} mechanism used in decay policies. In contrast to gated-V_{dd}, the drowsy mode preserves the state of the cache lines and results in a much smaller penalty when accessing deactivated, drowsy, lines. However, it is not without disadvantages: it does not save as much leakage as completely cutting off the power supply to the cache lines, and reduces reliability by making the memory cells more susceptible to soft errors. These two characteristics make for interesting comparisons between the two approaches and even more interesting hybrid schemes employing *both* approaches.

Decay versus drowsy: Parikh, Zhang, Sankaranarayanan, Skadron, and Stan examined energy savings for L1 data caches for the drowsy and cache decay mechanisms [178]. Their work shows that non-state-preserving techniques can outperform state preserving ones under certain conditions. More specifically, for *fast* L2 caches (5–8 cycle latency), cache decay in the L1 is better in terms of both performance and energy savings than a drowsy L1.

For the drowsy cache, Parikh et al. abandon the Simple policy of periodically putting all cache lines in drowsy mode, in favor of the more sophisticated decay policy based on the generational behavior of cache lines. The drowsy cache is therefore a *decaying* cache but instead of using gated-V_{dd} to turn-off cache lines, these are put in drowsy mode when they are "decayed."[15] The result is that would-be decay-induced misses turn into simple wake-ups from the drowsy mode. Consequently, the decay interval for the drowsy mode does not need to be as conservative as in the gated-V_{dd} mode.

The comparison, thus, turns into finding a break-even point: how much should the (non-state-preserving) decay penalty cost in order for decay (which saves more leakage but incurs decay-induced misses) to outperform the drowsy mode. A slower L2 makes the decay penalty more costly, wiping out the advantage of saving more. But the break even point also changes with temperature: the benefit of L1 decay is much higher at higher temperatures where it can outperform a drowsy L1 even when backed by a relatively slow L2.

Decay + drowsy hybrids on different cache levels: In contrast, to their conclusions regarding the L1, Li, Kadayif, Tsai, Vijaykrishnan, Kandemir, Irwin, and Sivasubramaniam examine in detail state-preserving versus non-state-preserving strategies in the L2 and conclude that non-state-preserving decay policies *in the L2* do not perform well compared to drowsy policies [154]. Simply put, the penalty for an L2 decay-induced miss—which has to go to memory—is just too high. Obviously, one has to be very conservative in the L2 not to incur decay-induced misses. This means using long decay intervals. This, in turn, implies that an overly conservative decay will probably not outpace the drowsy mode in leakage savings enough to turn the balance in its favor. Nevertheless, Li et al. conclude that combining cache decay in the L1 with drowsy policies in the L2 yields the best results overall if one considers the whole cache hierarchy [154].

☞ *hybrid policies for the L1 and L2*: Mirroring the Speculative-II and Speculative-IV decay policies discussed in Section 5.2.4, the Speculative-I and Speculative-III policies also deactivate L2 lines when they are transferred to the L1 but—this time—using the state-preserving drowsy mode instead of the gated-V_{dd} [154]. These two drowsy L2 policies are shown in Figure 5.15 (see also Figure 5.11 for a comparison with the gated-V_{dd} version). From the discussion so far, it is not surprising that Speculative-I turns out to be the best-performing policy in the L2: it saves considerable leakage while hurting performance very little.

On top of these two L2 policies, Li et al. examine more policies for the combined two-level cache hierarchy. Overall, the best policy in terms of EDP for the whole hierarchy

---

[15]As a matter of terminology, "*decay cache*" denotes the non-state-preserving version of decay while "*drowsy cache*" means the state-preserving (drowsy mode) form. The change in drowsy caches from the Simple policy to the generational policy is also followed in subsequent work, attempting to level the field between the two approaches by utilizing the same policy and emphasizing only the state-preserving (gated-V_{dd}) versus non-state-preserving (DVS) aspect.

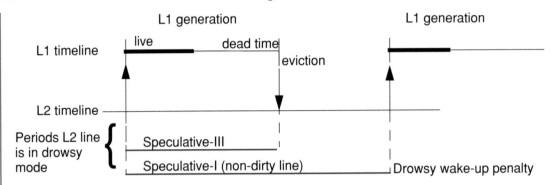

FIGURE 5.15: Time diagrams for the non-inclusion, state-preserving, drowsy policies. Speculative-I puts the L2 line in drowsy mode as soon as it is transferred to the L1. The L2 line remains in drowsy mode until the next L2 access (provided that there was no writeback of a dirty L1 line). The difference in Speculative-III is that the L2 line is awaken as soon as the L1 line is evicted (even if it is clean). Adapted from [154].

is a state-destroying decay in the L1 coupled with either state-preserving decay in the L2 or with the Speculative-I. State-destroying decay at both cache levels, although saving the most leakage, does not fare as well in EDP.

Decay+drowsy hybrid in the same cache: Zhang et al. were the first to use a hybrid state-destroying and state-preserving scheme for the instruction cache [246]. The control of the leakage mode is left to the compiler which chooses the state-preserving mode for cache lines containing instructions that are expected to appear again in the future and the state-destroying mode for instructions that are not coming back.

In the data cache, Meng, Sherwood, and Kastner take the hybrid idea one step further and combine decay and drowsy techniques, not on different cache levels but, in the same cache. Their goal is to examine the limits of leakage reduction assuming oracle knowledge—full knowledge of the access stream. They demonstrate optimal results for state-destroying (gated-V_{dd}) decay and the drowsy mode in the instruction and the data L1. Interestingly, their study shows that a hybrid scheme combining the decay and drowsy approaches yields the best results in a single cache [164].

The hybrid scheme is quite simple: it starts by putting a cache line in drowsy mode after a very short period of inactivity and proceeds to turn off the line completely if this period of inactivity persists beyond a larger "decay interval." This is as close to the optimal as a practical approach can get since it attacks the problems of decay and drowsy mode in a complementary fashion.

The drowsy mode problem of not saving enough leakage is solved by eventually completely switching off the cache line. Whereas, the two main decay problems are solved as follows. First, the problem of decay-induced misses is solved by waiting for a long time to make sure that a cache line is dead. Second, the problem of lost opportunity to save leakage waiting for the decay interval to pass (in high-leakage mode) is solved by using the drowsy mode while waiting. The hybrid approach works well assuming a constant temperature. Its trade-offs, however, change as temperature affects leakage.

5.3.4   Temperature

Although the dependence of leakage to temperature is not accounted for in the work of Meng et al., clearly the trade-off between drowsy and decay modes in a hybrid scheme cannot be static. The higher the temperature the more valuable the decay becomes, because it saves more. In contrast, at low temperatures, what matters most is to minimize dynamic energy penalty and performance loss; this argues in favor of the drowsy mode.

This temperature-driven trade-off is explored in a temperature-adaptive scheme [129]. The leakage control mechanism is a hybrid drowsy + decay scheme. The decision on how long to wait to enter the decay mode (i.e., how long is the decay interval) depends on the relative strength of the leakage power to dynamic power.[16] The dynamic power component is affected by the number of decay-induced misses which is a function of the decay interval.

At high temperatures, where leakage is relatively strong, the decay mode can be aggressively engaged since, even with an increase in dynamic power overhead, it maximizes the overall power savings. But when leakage currents are weak, the dynamic power overhead can dominate, regardless of the amount of saved leakage.

Temperature-driven adaptation consists of using a timing mechanism to stretch the decay interval at lower temperatures. This diminishes the leakage reduction from the decay mode but also minimizes its dynamic power overhead, leading to an overall reduction in power over a wide range of temperatures. The drowsy interval is not changed and is fixed at all temperatures.

While there are several ways to control the decay interval at various temperatures (for example, using a thermal sensor to set the global counter of a hierarchical counter mechanism), adaptation is based on the decay of quasi-static 4T memory cell. The decay interval is measured as the time it takes leakage to discharge a 4T cell. In other words, the strength of the leakage itself determines the decay interval. The key in this idea is that the discharge characteristic of a 4T memory cell, at various temperatures is ideally suited to be used as the decay interval. The resulting scheme consistently outperforms the decay mode or the drowsy mode alone and approaches the performance of a hybrid scheme with oracle decay interval selection [129].

---

[16]This technique assumes that temperature alone is the factor that changes the relative strength of leakage at runtime but dynamic scaling of the threshold voltage and the supply voltage could also have major effects.

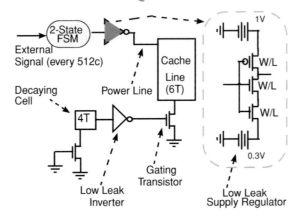

FIGURE 5.16: Decaying 4T timer for a hybrid decay + drowsy policy. Reproduced from [129]. Copyright 2005 IEEE.

☞ *decaying 4T timers*: A 4T DRAM cell can be used as a cheap temperature-sensitive timer [128]. The time it takes to discharge because of leakage is a measure of temperature. The timer in the temperature-adaptive hybrid technique is based on the same idea: the decay interval of each cache line is regulated by the decay of a 4T cell.

A simplified timing mechanism that sets a cache line in drowsy or decay mode is depicted in Figure 5.16. Implemented adjacent to each cache line, this mechanism adapts the decay interval to the temperature of the line's immediate surrounding area. The drowsy interval is fixed and is counted by a single-bit local counter. Accessing a line charges the 4T cell—i.e., resets the timer. As long as the 4T holds a "1" the cache line is connected to ground. If, however, the 4T is left unaccessed for a long period, it decays and gates the cache line via a low leak sleep transistor. As soon as the line is accessed again, the 4T reinstates the connection to ground. Additional hardware is needed to synchronize the decay of the 4T with the clock and cache operations.

5.3.5 Reliability

Finally, the different leakage-reduction techniques can have a very different reliability behavior compared to a standard cache. Degalahal, Vijaykrishnan, Kandemir, and Irwin examine cache decay and drowsy caches from a reliability perspective [63] and conclude that decay enhances reliability, whereas the drowsy mode makes caches more susceptible to soft errors.

Technology scaling fundamentally affects the reliability of devices, increasing the rate of soft (transient) errors. Such errors are primarily caused by external radiation such as alpha particle or high-energy neutron strikes but it is the physical characteristics of the devices that determine whether transient errors occur during these strikes.

Soft errors occur when a particle strike at a circuit node generates enough charge to cause a bit flip at that node. This happens when the collected charge from the strike exceeds a quantity known as $Q_{critical}$, which is proportional to the node capacitance and the supply voltage. The *Soft Error Rate (SER)* is exponentially dependent on $Q_{critical}$ and proportional to the cross-section area of the node (CS) and the environmental radiation flux (N_{flux}):

$$SER = N_{flux}/ \times CS \times e^{\left(\frac{-Q_{critical}}{Q_s}\right)}.$$

The exponential dependence to $Q_{critical}$ translates to an exponential dependence on the supply voltage. This means that DVS techniques such as the drowsy technique have a dramatic effect on reliability—increasing by at least an order of magnitude the SER. In contrast, cache decay improves reliability but invalidating a significant portion of the data, thus, reducing their exposure to soft errors. In addition, the early writeback induced by decay, helps protect dirty data by writing them back to the memory system.

Degalahal et al. further exploit decay's early writeback property in an adaptive error-protection scheme. The idea is to protect differently clean and dirty data, assuming that an error on clean data can be corrected by re-fetching the data from a lower level of the hierarchy. This leads to 11% savings in the dynamic energy expended on error protection in the L1.

A related technique to enhance reliability was proposed by Zhang, Gurumurthi, Kandemir, and Sivasubramaniam [245]. The technique, called *In-Cache Replication (ICR)*, is based on cache decay and replicates live cache lines in the dead space of the cache. Space vacated by dead lines is reclaimed and used to hold replicas of live lines. Replication can be performed vertically, across sets, by making a copy at a fixed distance from the original's set, or within the same set by reclaiming empty associative ways. Zhang et al. examine a number of variations of this idea and show that reliability can be significantly enhanced without compromising the performance [245].

5.3.6   Compiler Approaches for Decay and Drowsy Mode

All approaches to control leakage in caches, thus far, are based on hardware monitoring (e.g., decay) or even simple hardware techniques (e.g., the Simple policy for the drowsy mode). Compilers, however, can play an important role with the help of leakage control instructions.

Compiler involvement, assisted by profiling, has been initially proposed along with the hardware implementation of cache decay [127]. The compiler approach assumes the availability of instructions that access cache lines *and* turn them off immediately after. The idea is to find what instructions are *last-use* instructions for cache lines and replace them with leakage-control instructions. Unfortunately, it is difficult for the compiler to do a thorough analysis of the data without the help of run-time information. Profiling is necessary for this approach to work but

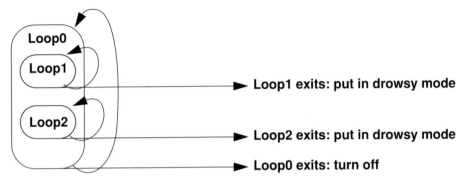

FIGURE 5.17: Compiler-controlled leakage management. Adapted from [246].

this is also its Achilles' heel. Different input sets, different data layouts, alter the usefulness of the profiling information. Profiling runs are also not desirable in many cases.

The situation, however, is more interesting when it comes to instruction caches. Compilers have a better ability of understanding and handling code. Through compile-time analysis, code can be enhanced with the ability to manage its leakage footprint in the instruction cache. For example, a loop can shut itself off (i.e., its footprint in the instruction cache) once it exits.

Zhang, Hu, Degalahal, Kandemir, Vijaykrishnan, and Irwin first proposed compiler leakage management for code [246]. They assume that loop instructions, as a group, can be selectively put into a low leakage mode (either state-destroying or state-preserving). Regardless of what happens to state this is referred to as *"turning off"* the instructions or cache lines. Finding and turning off *just* the cache lines belonging to a specific loop is bound to cost energy whether it is implemented in hardware or software. A simpler solution is proposed instead: *all* cache lines of the cache are turned off, blindly, upon exiting a loop.

The compiler can follow two strategies to turn off loops. The first is a *conservative* strategy in which instructions are turned off only if it is clear that they will not be accessed again. The second is an *optimistic* strategy in which instructions are turned off even if they are accessed in the future, as long as this is far enough in the future to produce a net gain.

In the example in Figure 5.17, the compiler cannot turn off loop1 or loop2 instructions under the conservative policy since it knows that they will be used again as part of loop0's execution. When loop0 exits, however, all code can be turned off. Under the optimistic strategy, both loop1 and loop2 can be turned off since their execution is interleaved. This strategy assumes that each one lasts long enough to warrant turning off the other.

The conservative strategy is combined with a state-destroying mechanism (e.g., gated-V_{dd}) since it only takes action when it knows with certainty that instructions (loops) are dead. Similarly, the optimistic strategy is combined with a state-preserving mechanism (e.g., drowsy mode) since it expects that eventually instructions are going to be reused. While the other two combinations of strategies and mechanisms are also possible (i.e., conservative with drowsy and optimistic with gated-V_{dd}), they are not as effective.

Figure 5.17 shows the behavior of a combined *hybrid* strategy where the optimistic and conservative strategies are used together. The hybrid strategy performs the best among all the other complier strategies. It also outperforms both the (state-destroying) decay and the drowsy techniques (with a fixed 4Kc decay interval) in terms of EDP in eight out of ten programs [246].

5.4   ARCHITECTURAL TECHNIQUES BASED ON V_T

The root of the exponential increase in subthreshold leakage can be traced to the lowering of the threshold voltage, V_T, as a result of technology scaling. Lowering the V_T is a matter of performance: device speed depends on the difference between the supply voltage, V_{dd}, and the threshold voltage V_T (see Section 5.1.1). This leads to a classic trade-off between speed and power that can be exploited to reduce the overall leakage consumption.

There are two broad approaches that exploit this trade-off, depending on whether V_T can be manipulated *dynamically* or is set at the design or manufacturing stage. In both cases, a low V_T yields increased speed at the expense of higher power consumption, whereas a high V_T reduces power consumption but also speed.

For the *dynamic* V_T technologies the most interesting tradeoff happens in relation to dynamic voltage scaling. Both V_{dd} and V_T scaling affect the device speed. V_{dd} scaling reduces dynamic power, whereas (upwards) V_T scaling reduces leakage power. The question then becomes, what to scale and how much for a given frequency of operation. This depends on the relative strength of the dynamic versus leakage power. Multiple studies have shown that one has to simultaneously scale *both* quantities to achieve the best possible results.

For the static, multiple-V_T technologies, such as MTCMOS, two popular design methodologies are generally followed.

- The first relies on transistor stacking: a block of low-V_T (high-leakage, high-speed) transistors are stacked with a high-V_T (low-leakage, low-speed) transistor. The high-V_T transistor is inserted between the high-speed logic block and one of the power rails (commonly the ground rail). When it is turned off, it disconnects the logic block from power, virtually eliminating leakage.

- The second populates non-critical paths of a circuit with high-V_T, low-leakage, transistors if there is enough slack to accommodate the increased latency [69]. Finally, in memory designs, asymmetric cells using high-V_T and low-V_T devices can save leakage depending on the stored bits [17].

5.4.1   Dynamic Approaches

Dynamically varying the threshold voltage of a device is accomplished by biasing the substrate body of a transistor. Two approaches have been proposed for biasing. In the first approach, called Reverse Body Bias (RBB), the substrate body of a fast, high-leakage, low-V_T device is biased to increase its V_T [132]. This results in dramatic reduction in leakage but slows down switching. The second technique, called Forward Body Bias (FBB), starts from the opposite direction. It takes a high-V_T, low-leakage, slow device and by applying the opposite bias reduces its V_T to make it fast but high leakage [13]. Adaptive approaches based on these two techniques have also been proposed, such as the Adaptive Body Bias (ABB) technique which uses reverse biasing.

> ☞   *a comparison of RBB to other leakage-reduction mechanisms*: At the circuit level, Chatterjee et al. [49] compare RBB to several techniques for subthreshold leakage reduction including transistor stacking [184], and dynamic voltage scaling (drowsy mode or DVS) [77]. They show that the stacking effect is the most effective means to reduce leakage power, but because it lowers the active current in the normal operation mode, it is also slow. They also show that lowering the supply voltage (DVS) is inferior in terms of both leakage savings and speed (low voltage also underrates I_{on} significantly). Thus, they conclude that RBB is the best compromise between leakage savings and speed in normal operation mode.
>
> RBB, however, requires the generation and routing of extra power supply to the body and well terminals of n- and p-MOS transistors. In addition, it requires the usage of a triple-well bulk CMOS process increasing the overall implementation cost [49]. If the ease of fabrication is taken into account, the best compromise in leakage savings and switching speed between high-leak and low-leak modes is the drowsy mode (DVS)—for state-preserving techniques—or gated-V_{dd} (V_{ss})—for non-state-preserving techniques.

Combined DVFS and ABB: Adaptive RBB techniques increase the threshold voltage and thus bring an exponential reduction in leakage power. However, the increase in threshold voltage reduces the ($V_{dd} - V_T$) difference, slowing down switching [195]:

$$\text{Delay} \propto \frac{V_{dd}}{(V_{dd} - V_T)^a}.$$

It is evident that either scaling V_{dd} or increasing V_T slows down switching. The question is which one to change for a given performance level.

- Voltage scaling, discussed extensively in Chapter 3, decreases dynamic power quadratically to the supply voltage:

$$P = C_{eff} \times f \times V_{dd}^2.$$

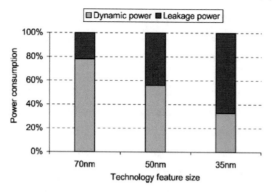

FIGURE 5.18: Relative contribution of dynamic and leakage power in an embedded processor. Reproduced from [231]. Copyright 2005 IEEE.

- Increasing V_T exponentially decreases the subthreshold leakage current and consequently subthreshold leakage power, in accordance to the formulas given in Section 5.1.1. Assuming $V_{gs} = 0$ and $V_{ds} = V_{dd}$, the formula for the subthreshold leakage power becomes:

$$P_{sub} = V_{dd} \, I_{Dsub} = V_{dd} \, I_{s0} \left(1 - e^{\frac{-V_{dd}}{V_t}} \right) e^{\frac{-V_T - V_{off}}{n \, v_t}}.$$

Considering dynamic or leakage power independently, the performance can be traded for power by scaling either V_{dd} or V_T. Because in both cases performance degradation is linear to the scaling of the V_{dd} or V_T, whereas power savings are either quadratic or exponential, the resulting improvement in EDP is substantial.

Considering, however, *total* power, the sum of the dynamic and leakage powers, it is not obvious which quantity is more profitable to scale for a given performance degradation. This depends on the relative contribution of the two components to the total power consumption. For example, Yan, Luo, and Jha [231] consider the three scenarios, shown in Figure 5.18, for the relative contribution of dynamic and leakage power in an embedded processor.

In the 70 nm technology, scaling the supply voltage is bound to have a greater effect than raising V_T, for a given performance level—a given frequency—since dynamic power dominates in this technology. In contrast, in the 35 nm technology, increasing V_T is the more profitable route. Not only the balance of dynamic and leakage power shifts across technologies, or among different implementations in the same technology, but also changes dynamically as a function of temperature which has a profound effect on leakage. This aspect, however, has not been researched adequately.

For a given frequency—a given switching delay—the best possible power savings come from carefully adjusting *both* V_{dd} and V_T, depending on the balance of dynamic versus leakage power at that point. While the $V_{dd} - V_T$ difference determines switching speed, maximum gains in power consumption come from a combined adjustment of the two. Three independent studies come to the same conclusion.

- Duarte, Vijaykrishnan, Irwin, Kim, and McFarland study the impact of scaling on a number of approaches for dynamic power reduction [70]. Among their experiments they simultaneously scale the supply voltage (V_{dd}) and the body-to-source bias voltage (V_{bs}), i.e., they simultaneously perform DVS and ABB. Their study is not constrained in any variable, meaning that they examine a wide spectrum of possible values for the two quantities. Their results show a clear advantage over DVS alone.

- The work of Martin, Flautner, Mudge, and Blaauw combines DVS and Adaptive Body Biasing to lower both dynamic and static power of a microprocessor during execution [163]. They derive a closed-form formula for the total power consumption, expressing it as a function of V_{dd} and V_{bs} (the body-to-source bias controlled by ABB). The formula is the following:

$$ P = C_{\text{eff}} \, V_{dd}^2 \, f + V_{dd} \, K_3 \, e^{K_4 \, V_{dd}} \, e^{K_5 \, V_{bs}} + |V_{bs}| \cdot I_j, $$

where, K_3, K_4, and K_5 are constants derived from simplifications made by expressing V_T as a function of V_{dd} and V_{bs}. The formula also includes a term for junction leakage (I_j) which the authors consider important in this situation. In a similar manner, they arrive at the following formula for the frequency, f:

$$ f = (L_d \, K_6)^{-1} \, ((1 + K_1) \, V_{dd} + K_2 \, V_{bs} - V_{th1})^a, $$

where K_1, K_2, K_6, and V_{th1}, are constants derived from approximations, L_d is the depth of the logic path in relation to an inverter, and a is the exponent of the alpha-power delay model of an inverter (here $a = 1$). As it is evident from the above formula, f—performance—is a linear function of V_{dd} and V_{bs}.

Martin et al. use the system-level technique of automatic performance setting presented in Chapter 3 as a DVS-only technique (Section 3.2). In this technique, deadlines are derived from monitoring system calls and interprocess communication. The performance setting algorithm sets the processor frequency for the executing workload so as to not disturb its real-time behavior.

Solving the system of the two equations above for a given performance setting, Martin et al. are able to estimate the most profitable combination of V_{dd} and V_{bs} to maximize power consumption savings. The approach can deliver savings *over* DVS alone of 23% in a 180 nm process and 39% in a (predicted) 70 nm process [163].

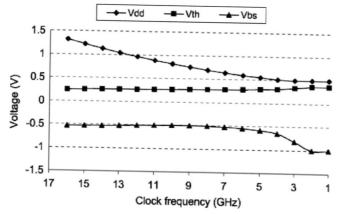

FIGURE 5.19: Combined DVS and ABB for an embedded processor with hard real-time constraints. Reproduced from [231]. Copyright 2005 IEEE.

- Finally, Yan, Luo, and Jha study the application of combined DVS and ABB in heterogeneous distributed real-time embedded systems [231]. In analogy to the work of Martin et al., they determine the lowest frequency of operation that can satisfy the real-time constraints of an embedded system using the worst-case analysis. In contrast to the previous work, the deadlines are known and are hard real time.

  Given the required operation frequencies, Yan et al. show that both V_{dd} and V_T have to scale to obtain the minimum power across the range of frequencies for a 70 nm technology (Figure 5.19). Initially, V_{dd} is scaled as this yields the most power reduction for the given frequencies. However, at some point V_{dd} scaling reduces dynamic power to the point where leakage starts to dominate, thus triggering ABB (scaling the body-to-source bias V_{bs}). Yan et al. conclude their study by presenting a heuristic on-line algorithm that can estimate the proper V_{dd} and V_{bs} for a list of tasks with a slack in their execution schedule.

5.4.2   Static Approaches

In the static, multiple-V_T approaches, transistors on the same die are manufactured as either high-V_T or low-V_T (by varying the channel-doping profile, or the oxide thickness, or the channel length). These are referred to as multi-threshold voltage CMOS (MTCMOS) technologies with the appropriate libraries providing devices with different V_T. The designer or architect selects where the high-V_T or low-V_T devices go at design time and the chip is manufactured accordingly. There are two major design methodologies in using multiple threshold voltages.

- The first methodology is to use *low-V_T* devices (high-performance, high-leakage) in the performance-critical paths of a design and use *high-V_T* (lower-performance, low-leakage) everywhere else; as long as the increased delay of the high-V_T devices can be hidden in the slack this saves leakage without affecting speed. An example of this technique is in dynamic domino logic [124] which is described in detail in Section 5.4.3. The asymmetric cell design presented in Section 5.4.4 is an analogous methodology but is intended for memory and exploits a different storage asymmetry [17, 18].

- The second design methodology uses *high-V_T* sleep transistors to gate off the power to high-speed, low-V_T, logic when such a logic is idle. This is similar to the gated-V_{dd} approach mentioned above, the difference being the use of dual threshold voltages. The leakage reduction in the low-V_T logic is still a result of the stacking effect as described previously, but now enhanced by the presence of the *high-V_T* sleep transistors. We will not expand separately on this methodology since it is a dual-V_T version of the gated-V_{dd} approach already described in Section 5.2.

5.4.3   Dual-V_T in Functional Units

One of the few architectural-level studies for the use of dual-V_T in functional units is by Dropsho, Kursun, Albonesi, Dwarkadas, and Friedman [69]. For performance reasons, functional units are typically designed using dynamic domino logic instead of static CMOS. Figure 5.20 contrasts an AND gate in static CMOS and in dynamic domino logic. In terms of dynamic power the difference in the two designs is that, every clock cycle, the domino logic is charged and discharged (if needed) by the evaluation of its inputs. As mentioned in Chapter 4, Section 4.2, this can consume power *even when the inputs do not change*. The only way out is to stop the clock, i.e., clock gate the entire circuit not just its inputs. With respect to static power,

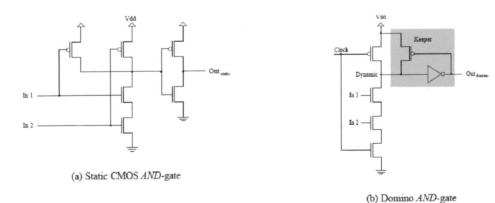

(a) Static CMOS *AND*-gate

(b) Domino *AND*-gate

FIGURE 5.20: AND gate in static CMOS and in dynamic domino logic. The "Dynamic" node in (b) is precharged during the low phase of the clock. Inputs are evaluated during the high phase of the clock and can discharge the dynamic node. Reproduced from [69]. Copyright 2002 IEEE.

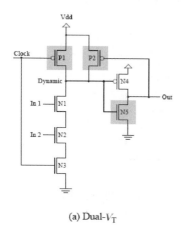

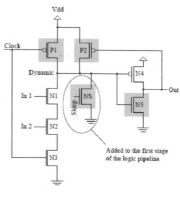

(a) Dual-V_T

(b) Dual-V_T with sleep mode

FIGURE 5.21: Low-leakage dual-V_T AND gate (a) and same with sleep mode (b). In (a) P1, P2, and N5 are high-V_T devices (they are not on the critical path for evaluation). In (b) the NS high-V_T transistor is the "sleep" transistor that discharges the dynamic node. Reproduced from [69]. Copyright 2002 IEEE.

leakage paths in dynamic domino logic depend on the state of the internal dynamic nodes. This property is exploited for the implementation of a sleep mode specific to domino logic.

The low-leakage domino logic is built using dual-V_T devices according to the MTCMOS design methodology described above. By selectively using high-V_T devices in the non-critical paths, performance is not compromised. Figure 5.21 shows the integration of high-V_T devices (shaded transistors) in the domino-logic AND gate. Evaluation speed is not affected by this configuration.

However, an asymmetry in the leakage currents is created. This asymmetry depends on the voltage level at the internal dynamic node. If either input is low, the dynamic node remains charged, resulting in a large subthreshold leakage current through the high-leakage transistors N1, N2, N3, and N4. But, when the dynamic node is discharged, the low leakage transistors P1, P2, and N5 are strongly cutoff and the leakage in the whole circuit is dramatically reduced.

Thus, to put a dynamic domino circuit in a low-leakage state, all that is needed is to discharge its dynamic nodes (and of course clock-gate it to prevent them from getting recharged). This can only be done if the circuit is idle since in the low-leakage state it cannot evaluate its inputs—it first needs to be recharged. This is a distinct technique, for a low-leakage state, from the power gating techniques presented in Section 5.2.6 or from ordinary clock gating for dynamic power.

Similarly to switching on and off the sleep transistor in the power-gating techniques (Section 5.2.6), discharging the dynamic nodes of a domino circuit carries an energy penalty. The energy cost is proportional to the number of nodes discharged to enter the sleep mode. As

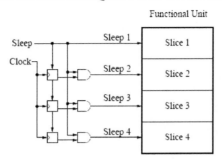

FIGURE 5.22: Gradual sleep mode. Reproduced from [69]. Copyright 2002 IEEE.

with power gating, the question then becomes, whether the idle time of the functional units is long enough for the gains in leakage energy to outpace the penalty of entering the low-leakage mode. Typically, it may not be profitable to enter in low-leakage mode for very short idle periods.

Unfortunately, the distribution of the idle periods for the integer functional units is dominated by short times [69]. This means that an overly aggressive policy to enter the sleep mode is probably not optimal. For this reason, Dropsho et al. propose a *gradual sleep* policy that puts the functional unit in sleep mode in stages. The technique is shown in Figure 5.22. The functional unit is divided in *slices* which are put in sleep mode consecutively as long as the functional unit remains idle. As soon as it is needed again, all slices are brought back to active mode and are precharged.

The gradual sleep technique can tie very well with the narrow-width operand techniques discussed in Chapter 4, Section 4.3, in relation to dynamic power. However, such a possibility was not considered. Nevertheless, the gradual sleep technique effectively harvests the idle times of the functional units for leakage savings without incurring large penalties, thus, making it very difficult for a more sophisticated technique to do much better.

5.4.4   Asymmetric Memory Cells

Using dual-V_T in functional units relies on timing asymmetries in the circuits. Non-critical paths can be populated by slower, high-V_T transistors taking up some of the time slack. Memory structures, however, are regular when it comes to timing. Azizi, Moshovos, and Najm, exploit a different kind of asymmetry, a *storage asymmetry*, to selectively introduce high V_T devices in SRAM design [17, 18].

Their technique relies on an imbalance in the number of ones and zeros that appears in the memory system for ordinary programs. The observation is that the number of zeros by

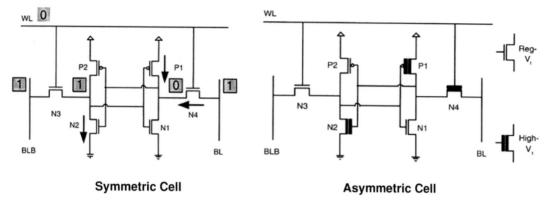

Symmetric Cell   **Asymmetric Cell**

FIGURE 5.23: Symmetric versus asymmetric SRAM cell. Voltage levels (shaded 1 and 0) and leakage paths are shown in the ordinary symmetric cell when it is holding a "0." The Asymmetric Cell utilizes high-V_T (low-leakage but slow) transistors for N2, P1, and N4 (the leakage paths). The result is low leakage in state 0 (reduced by 40°) but also higher latencies in discharging the bitlines (46% for the BL and 12% for the BLB). Reproduced from [18]. Copyright 2003 IEEE.

far outstrips the number of ones in memory. And this holds for both data and instructions, regardless of whether they are *actively* or rarely accessed.

This observation implies that the number of zero bits in the cache is also proportionally greater than the number of ones. We have encountered a similar approach for dynamic power in Chapter 4, Section 4.4.1, but at the *byte* level. Villa, Zhang, and Asanović observe that a large percentage of bytes in the memory system are zero, and therefore can be represented by a single bit, the Zero Indicator Bit (ZIB) [221]. Although they operate different granularity and for different kinds of power (dynamic versus leakage), the Azizi et al. and the Villa et al. techniques are closely related.

The idea in the Azizi et al. technique is to selectively use high-V_T devices to build a memory cell that is more power-efficient in holding zeros than ones. Such a cell is shown in Figure 5.23 (on the right) and contrasted to a standard cell (on the left). The transistors on the leakage paths (when the standard cell holds a zero) are replaced with high-V_T devices. The resulting asymmetry saves 40× of leakage when holding a zero—holding a one makes no difference. At the same time, an asymmetry is created in the access times: the discharge of the bit-line (BL) is slowed down by 46%, whereas the discharge of the negated bit-line (BLB) is slowed much less, only 12%. In contrast, building a whole cell out of high-V_T transistors saves leakage no matter what the stored value but the access speed is symmetrically impaired in both bitlines (BL and BLB).

The asymmetry in the access times proves to be the key in making the scheme perform on a par with ordinary cells. With a novel sense-amp design, complete with dummy bitlines

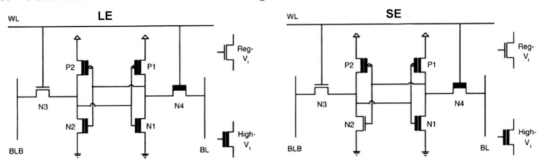

FIGURE 5.24: Two improved Asymmetric Cells: the Leakage Enhanced (LE) and Speed Enhanced (SE). Reproduced from [18]. Copyright 2003 IEEE.

connected on extra dummy cells, Azizi et al. are able to exploit the timing asymmetry in the discharge of the bitlines and trick the sense-amp in detecting zeros almost as fast as detecting ones. The details are explained in [17, 18].

The asymmetric design in Figure 5.23 is only one of the nine plausible asymmetric designs—but not the best. Azizi et al. describe and evaluate the other nine designs and determine for each the leakage savings when storing a zero or a one and the respective latencies in the discharge of the bitlines BL and BLB [18].

Two improved designs are shown in Figure 5.24. The first is a leakage enhanced (LE) version and the second is a speed enhanced (SE) version. In the LE cell, the replacement of N1 and P2 by high-V_T devices improves the leakage power consumption when the cell holds a 1. The savings are 40% and 7% for holding a 0 or a 1 respectively. The discharge of the BLB is slowed further, reaching 61% of the normal low-V_T cell but the discharge of the BL is not affected remaining 12% slower than the normal cell. Still the extra latency for the BLB is hidden by the new sense-amp design that detects a zero from the dummy bitlines exactly because the BLB discharge is slowed. The SE design puts the emphasis on speed. Although its leakage reduction is not as impressive (only 2× and 7× compared to a normal cell when holding a 0 and a 1 respectively), its response time, with the help of the new sense-amp, is virtually identical to a normal (low-V_T) cell.

CHAPTER 6

Conclusions

While performance considerations have always played a prominent role in computer architecture, power consumption has been an increasingly significant driver of the field in recent years. Both the increase in dynamic power and even more so the exponential increase of static power have brought sweeping changes in the way we design and build processors.

In the last two decades or more, the industry has benefited from exponential performance improvments stemming from a combination of exponential increase in clock frequency, along with architectural innovations. Clock frequencies today, however, have hit the power wall. They are currently increasing at a much slower pace than ever before, and in some cases have *decreased* from prior chip generations. Instead of frequency increases as the main performance driver, other options become paramount. In particular, there is much greater focus today on chip multiprocessors (CMPs) and on application-specific designs. Both of these have interesting and important power implications, as we will discuss below. In the subsections that follow, we conclude this book with summaries of current state of the art, as well as views of promising future techniques and trends.

6.1 DYNAMIC POWER MANAGEMENT VIA VOLTAGE AND FREQUENCY ADJUSTMENT: STATUS AND FUTURE TRENDS

Much of the early focus on dynamic power management concerned possible voltage and frequency adjustments, and a wide range of dynamic or offline policies to guide them. This work has had high leverage both in terms of published research results as well as in terms of commercial adoption. Part of the success of these techniques has stemmed from the quadratic (or more) influence of (V, f) scaling on power; it has unequivocally been a high-payoff technique. Another reason for its success stems from the fact that the *mechanism* of DVFS has been built into many processors, and its existence has given researchers an excellent platform with which to expeiriment with increasingly effective *policies* for managing it.

Looking forward, it seems unlikely that reliance on further DVFS techniques will offer as significant further payoffs in the future. First, much of the "low-hanging fruit" has been harvested; further success will likely result in smaller improvements to existing policies. Second, as leakage energy plays a greater role in total power dissipation, DVFS techniques are less promising because of their detrimental impact on leakage.

Third, and most importantly, while DVFS has been the backbone of power management in most processors, it faces some important limitations that will diminish its value in future processors. While DVFS offered quadratic reduction in power ($\sim V^2 f$) with only linear slowdown ($\sim f$) its practical value rests on a wide range of supply voltages (e.g., 3 V to 1 V). This range is continuously shrinking with the lowering of the upper limit but not a corresponding lowering of the lower limit. ITRS projects for the upcoming generations (2007–2014) supply voltages in the range of 0.9 V to 0.6 V giving a useful dynamic range of 0.2 V–0.3 V. In addition, the lower supply voltages significantly increase the susceptibility of the devices to soft or transient errors. And this for technologies that are inherently unreliable to begin with! It is very likely that in future processors, DVFS will either be impractical or unfruitful if the power needed to correct errors outweighs the power saved by voltage scaling.

6.2   DYNAMIC POWER REDUCTIONS BASED ON EFFECTIVE CAPACITANCE AND ACTIVITY FACTOR: STATUS AND FUTURE TRENDS

A significant portion of both published research and industry practice is devoted in reducing the effective capacitance ($C_{\text{effective}} = A \times C$) term of the dynamic power equation. Classifying the types of excess switching activity can offer a summarized checklist of steps for architects interested in power-efficient design:

- *Check for idle-unit switching.* Clock gate every unit that can be deterministically ascertained as idle. The granularity for clock gating should be such that its overhead is insignificant.

- *Check for idle-width.* Check to see if the *width* of various structures, functional units, data paths, memories (latches, registers, caches, etc.) is too wide for the common case. If it is, either (dynamically) disable the unused width or overload the width by packing more than one "narrow" operation at a time.

- *Check for idle-capacity.* Check to see if the *capacity* of large structures is fully utilized. If not, *resize* to accommodate the common case without inordinately harming performance. This eliminates switching in the unused parts of a large structure. It is also beneficial for static power if the disabled parts, after a resizing, are completely powered down or alternatively put into a state-preserving low-leakage mode.

- *Check for parallel/speculative activity.* Parallel speculative activities such as set-associative cache access, parallel searches, indiscriminate snooping, etc., can be significantly reduced by performing first only what is most likely to succeed.

- *Check for repetitive/cacheable activity.* Complex computation that repeats verbatim can be possibly memoized—that is, stored as an association of the inputs to a specific output. If retrieving the result consumes less energy than computing it, this method can yield excellent improvements in power. Caches can also be cached. After all, the cache hierarchy itself is a power optimization.

- *Check for Speculative activity* at a large scale that is wasted on misspeculations. Find a way to reduce the work performed during probable misspeculations.

- *Check for value-dependent activity.* A different encoding can sometimes lead to a totally different switching profile. Although we present this type of activity last, an encoding change is a fundamental and fairly low-level optimization that can performed before any of the above optimizations.

Most of low-power research and practice centers around these types of *excess or avoidable* activity. As long as there is a consistent and persistent effort to systematically address each and every one of these cases, the power inefficiency that we have seen, especially in out-of-order superscalar processors, can be largely rectified.

Optimizing a component to reduce excess activity in some cases may have undesired consequences. In these situations, a *work steering* strategy can be used where both the optimized and the unoptimized versions of the component are provided; work is then dynamically steered to the appropriate component according to run-time conditions.

Looking forward, one can view future CMP trends through the prism of their impact on power. In particular, the effective capacitance clearly demonstrates the inefficiency of the ILP "uni-core" approach. With each new and more complex ILP unicore, the core size has increased, leading to longer wires on average. In addition, the activity factor also increases. While performance has increased somewhat in return the marginal performance benefits have been decreasing with each generation, while power and power density have been increasing.

In other words, each successive ILP architecture was less power-efficient for its performance gains. The move to multi-core architectures, using a power-efficient core as a building block, has many benefits. Among them, it allows the activity factor to be controlled on a per-core basis, and allows average wire lengths to be primarily limited by core size, rather than die size.

While multi-cores reduce global wiring, they do not eliminate it. In particular, on-chip processors are interconnected either by a shared bus or by some sort of on-chip (perhaps packet-switched) network-on-chip (NoC). Interesting future research directions must focus on power-efficient implementations of these interconnects, as well as optimization of the inter-processor communication patterns in order to save power and energy.

6.3 LEAKAGE POWER REDUCTIONS: STATUS AND FUTURE TRENDS

Leakage energy remains the most vexing of the problems raised in this book. Below the architecture level, process technology has come up with solutions to reduce gate leakage (e.g., high-k dielectrics) but subthreshold leakage remains a problem. Circuit techniques to "power-gate" regions of a design are promising, but continue to require increases in wiring or in cicuit complexity. Furthermore, in contrast to dynamic power, there have been fewer good abstractions for how architects can reason about leakage energy and control it, without dropping to circuit-level design.

Currently and into the future, the leakage problem is exacerbated by high-performance chips operating at or near the top of their thermal envelope. Significant effort must be expended at the architectural level to fight leakage even at the smallest structures to avoid thermal runaway effects because of the exponential relation of subthreshold leakage to temperature. Scaling the supply voltage does reduce both the dynamic power and leakage power (e.g., drowsy techniques) but, as we mentioned above, reduced supply voltages significantly compromise reliability of the upcoming nanoscale technologies. Thus, drowsy techniques are expected to cause significant increases in soft errors.

Overall, we see leakage management as one of the key areas of future architecture-level power research. A particularly pressing need is on general techniques and abstractions for reasoning about leakage in more circuit-oblivious ways.

6.4 FINAL SUMMARY

The past decade has seen a dramatic evolution in how architects view power dissipation and related issues. Roughly ten years ago when architecture conferences began publishing power-oriented papers, they were somewhat at the fringe of the community, with most attention still be focused on performance-enhancing/power-oblivious techniques. This was followed by years in which the "power problem" saw intense research interest and a flurry of interesting and varied results on how to build power-efficiency into a range of microprocessor architectures and structures.

We now appear to be at a new inflection point in the timeline of architectural power research. In particular, power now plays some role in almost every paper or technique proposed, rather than being a fringe issue in isolated papers. On the other hand, it is rare for power to be the sole focus, and we feel this is rightly so. Rather, power is viewed holistically as part of a family of design goals, along with performance, reliability, design verifiability, and others. Architects of the future will need to maintain this balanced and holistic view across the entire design space. Without this holistic approach, architects are wasting the leverage offered to us by being high-level guides for design trends, and by being the "translators" at the hardware software interface.

Glossary

This glossary provides short descriptions for *acronyms* appearing in this book.

- ALU: Arithmetic/Logic Unit.
- CAM: Content Addressable Memory, a fully-associative memory organization.
- CMP: Chip MultiProcessor, a multiprocessor comprising several interconnected central processing units (CPUs)—and possibly their caches—on the same chip.
- DVS: Dynamic Voltage Scaling.
- DVFS: Dynamic Voltage/Frequency Scaling.
- EDP: A power-efficiency metric taking into account the impact on both energy and performance.
- ED^2P: A power-efficiency metric taking into account the impact on both energy and performance but with an emphasis on performance.
- FIFO: First-In First-Out: a common organization for hardware queues.
- FU: Functional Unit.
- GALS: Globally Asynchronous – Locally Synchronous, refers to a design style where a processor or other complex device is composed of multiple units operating asynchronously using different clocks. Each unit is clocked with its own clock and it is (internally) synchronous.
- IPC: Instructions Per Cycle, a measure of performance.
- IQ: Instruction Queue, a queue of instructions waiting to be issued to execution units.
- IW: Instruction Window, the main hardware structures handling in-flight instructions in an out-of-order, super-scalar, CPU.
- L1, L2: Level-1 and Level-2 caches. If not explicitly indicated, a data cache (instead of an instruction cache) is assumed for the Level-1 and a unified cache (caching both data and instructions) is assumed for the Level-2.
- LRU: Least Recently Used: a common replacement algorithm for set-associative caches.
- LSQ: Load/Store Queue, a queue of load/store instructions awaiting address resolution or completion from the memory system.

- MCD: Multiple-Clock Domains, refers to a design style where more than one clock is available on a chip.
- MediaBench: An embedded benchmark suite by the MediaBench Consortium.
- MiBench: A commercially representative embedded benchmark suite from the University of Michigan.
- MRU: Most Recently Used: the last accessed item in an associative structure (or cache set).
- Multi-core: a more recent name for a CMP.
- OLDEN: A pointer-intensive benchmark suite.
- Perfect (Club): A supercomputer benchmark suite.
- SPEC95, SPEC2000, SPEC2K, SPEC-int, SPEC-fp: Benchmark suites developed by the Standard Performance Evaluation Corporation.
- SPLASH, SPLASH-2: Parallel benchmark suites compiled at Stanford University.
- SMT: Simultaneous MultiThreading, executing several threads concurrently in an out-of-order, super-scalar CPU by replicating the architectural state for each thread.
- SRAM: Static Random Access Memory.
- TLB: Translation Look-aside Buffer, a fully-associative structure holding recent virtual-to-physical address translations.
- uop(s), μop(s): Micro-operation(s), Intel's term for the internal instruction format in Pentium Pro and newer processors. In these processors the well known x86 or IA-32 Instruction Set Architecture (ISA) is translated internally to uops.

Bibliography

[1] J. Abella, A. González, X. Vera, and M. F. P. O'Boyle, "IATAC: a smart predictor to turn-off L2 cache lines," *ACM Trans. Architecture Code Optim.*, Vol. 2, No. 1, 2005.

[2] *Advanced Micro Devices Corp. 2002.* AMD-K6 Processor Mobile Tech Docs. http://www.amd.com.

[3] A. Agarwal and S. D. Pudar, "Column-associative caches: a technique for reducing the miss rate of direct mapped caches," in *Proc. 20th Int. Symp. on Computer Architecture (ISCA-20)*, pp. 179–190, IEEE, 1993.

[4] A. Agarwal, J. Hennesy, and M. Horowitz, "Cache performance of operating systems and multiprogramming," in *ACM Transactions on Computer Systems*, pp. 393–431, Nov. 1988. doi:10.1145/48012.48037

[5] H. Akkary, R. Rajwar, and S. T. Srinivasan, "Checkpoint processing and recovery: Towards scalable large instruction window processors," in *Proc. 36th Annual IEEE/ACM Int. Symp. on Microarchitecture (MICRO-36)*, 2003.

[6] A. Alameldeen and D. Wood, "Adaptive cache compression for high-performance processors," in *Proc. 31st Int. Symp. on Computer Architecture (ISCA-31)*, 2004.

[7] D. H. Albonesi, "Dynamic IPC/clock rate optimization," in *Proc. 25th Int. Symp. on Computer Architecture (ISCA-25)*, 1998.

[8] D. H. Albonesi, "Selective cache ways: On-demand cache resource allocation," in *Proc. 32nd Annual IEEE/ACM Int. Symp. on Microarchitecture (MICRO-32)*, pp. 248–259, Nov. 1999.

[9] D. H. Albonesi, R. Balasubramonian, S. Dropsho, S. Dwarkadas, E. G. Friedman, M. C. Huang, V. Kursun, G. Magklis, M. L. Scott, G. Semeraro, P. Bose, A. Buyuktosunoglu, P. W. Cook, and S. Schuster, "Dynamically tuning processor resources with adaptive processing," *IEEE Computer*, Vol. 36, No. 12, pp. 49–58, 2003.

[10] M. Alidina, G. Burns, C. Holmqvist, E. Morgan, D. Rhodes, S. Simanapalli, and M. Thierbach, "DSP16000: A high performance, low-power dual-MAC DSP core for communications applications," in *Proceedings of the IEEE Custom Integrated Circuits Conference*, 1998.

[11] M. Alidina, J. Monteiro, S. Devadas, A. Ghosh, and M. Papefthymiou, "Precomputation-based sequential logic optimization for low power," *IEEE/ACM International Conference on Computer-Aided Design*, Nov 1994.

[12]   B. Allu and W. Zhang, "Static next sub-bank prediction for drowsy instruction cache," in *Proceedings of the 2004 International Conference on Compilers, Architecture, and Synthesis for Embedded Systems*, pp. 124–131, 2004. doi:10.1145/1023833.1023852

[13]   H. Ananthan, C. H. Kim, and K. Roy, "Larger-than-V_{dd} forward body bias in sub-0.5 V nanoscale CMOS," in *Proc. Int. Symp. on Low Power Electronics and Design (ISLPED)*, 2004.

[14]   M. Anis, S. Areibi, M. Mahmoud, and M. Elmasry, "Dynamic and leakage power reduction in MTCMOS circuits using an automated efficient gate clustering," in *Proc. Design Automation Conf. (DAC)*, 2002.

[15]   M. Annavaram, E. Grochowski, and J. P. Shen, "Mitigating Amdahl's law through EPI throttling", in *Proc. 32nd Int. Symp. on Computer Architecture (ISCA-32)*, pp. 298–309, 2005. doi:10.1109/ISCA.2005.36

[16]   J. L. Aragón, J. González and A. González, "Power-aware control speculation through selective throttling," in *Proc. 9th Int. Symp. on High-Performance Computer Architecture (HPCA-9)*, Feb. 2003.

[17]   N. Azizi, A. Moshovos, and F. N. Najm, "Low-leakage asymmetric-cell SRAM," in *Proc. Int. Symp. on Low Power Electronics and Design (ISLPED)*, 2002.

[18]   N. Azizi, F. N. Najm, and A. Moshovos, "Low-leakage asymmetric-cell SRAM" *IEEE Trans. Very Large Scale Integration (VLSI) Systems*, Vol. 11, No. 4, Aug. 2003.

[19]   R. I. Bahar and S. Manne, "Power and energy reduction via pipeline balancing," in *Proc. 28th Int. Symp. on Computer Architecture (ISCA-28)*, 2001.

[20]   R. I. Bahar, G. Albera, and S. Manne, "Power and performance tradeoffs using various caching strategies," in *Proc. Int. Symp. on Low-Power Electronics and Design*, 1998.

[21]   R. Balasubramonian, D. Albonesi, A. Buyuktosunoglu, and S. Dwarkadas, "Memory hierarchy reconfiguration for energy and performance ingeneral-purpose processor architectures," in *Proc. of the 33rd Annual IEEE/ACM Int. Symp. on Microarchitecture (MICRO-33)*, 2000.

[22]   K. Basu, A. Choudhary, J. Pisharath, and M. Kandemir, "Power protocol: reducing power dissipation on off-chip data buses," in *Proc. 35th Annual IEEE/ACM Int. Symp. on Microarchitecture (MICRO-35)*, 2002.

[23]   B. Batson and T. N. Vijaykumar, "Reactive-associative caches," in *Proceedings of the International Symposium on Parallel Architectures and Compiler Techniques (PACT)*, pp. 49–60, Aug. 2001. doi:10.1109/PACT.2001.953287

[24]   N. Bellas, I. Hajj, C. Polychronopoulos, and G. Stamoulis, "Energy and performance improvements in microprocessor design using a loop cache," in *International Conference on Computer Design (ICCD)*, 1999.

[25]   N. Bellas, I. Hajj, and C. Polychronopoulos, "Using dynamic cache management techniques to reduce energy in a high-performance processor," in *Proc. Int. Symp. on Low Power Electronics and Design (ISLPED)*, pp. 64–69, 1999.

[26]   L. Benini, A. Macii, E. Macii, M. Poncino, and R. Scarsi, "Synthesis of low-overhead interfaces for power-efficient communication over wide buses," in *Proc. ACM/IEEE Design Automation Conference (DAC)*, pp. 128–133, 1999.' doi:10.1145/309847.309898

[27]   L. Benini, G. Demicheli, E. Macii, D. Sciuto, and C. Silvano, "Asymptotic zero-transition activity encoding for address busses in low-power microprocessor-based systems," in *Proceedings of the Great Lakes Symposium on VLSI*, pp. 77–82, 1997. doi:10.1109/GLSV.1997.580414

[28]   L. Benini, G. De Micheli, E. Macii, D. Sciuto, and C. Silvano, "Address bus encoding techniques for system-level power optimization," in *Proc. Design, Automation and Test in Europe (DATE)*, pp. 861–866, Feb. 1998. doi:10.1109/DATE.1998.655959

[29]   J. Birnbaum and S. Williams, "Physics and the information revolution," *Physics Today*, Vol. 54, No. 1, pp. 38–42, Jan. 2000. doi:10.1063/1.882936

[30]   B. Bloom, "Space/time tradeoffs in hash coding with allowable errors," *Commun. ACM*, Vol. 13, No. 7, 1970. doi:10.1145/362686.362692

[31]   M. T. Bohr, R. S. Chau, T. Ghani, and K. Mistry, "The high-k solution," *IEEE Spectrum*, Oct. 2007.

[32]   S. Borkar, "Design challenges of technology scaling," *IEEE Micro*. Jul–Aug 1999. Vol. 19, No. 4, pp. 23–29, 1999. doi:10.1109/40.782564

[33]   S. Borkar, T. Karnik, S. Narendra, J. Tschanz, and A. Keshavarzi, and V. De, "Parameter variations and impact on circuits and microarchitecture," in *Proc. Design Automation Conference (DAC)*, June 2–6, 2003.

[34]   E. Brekelbaum, J. II Rupley, C. Wilkerson, and B. Black, "Hierarchical scheduling windows," in *Proc. 35th Annual IEEE/ACM Int. Symp. on Microarchitecture (MICRO-35)*, 2002.

[35]   D. M. Brooks, P. Bose, S. E. Schuster, H. Jacobson, P. N. Kudva, A. Buyuktosunoglu, J.-D. Wellman, V. Zyuban, M. Gupta, and P. W. Cook, "POWER-AWARE MICROARCHITECTURE: design and modeling challenges for next-generation microprocessors," *IEEE Micro*, Nov./Dec. 2000.

[36]   D. M. Brooks, P. Bose, V. Srinivasan, M. Gschwind, P. Emma, and M. Rosenfield, "New methodology for early-stage, microarchitecture-level power-performance analysis of microprocessors," *IBM J. Res. Dev.*, Nov./Dec. 2003.

[37]   D. Brooks and M. Martonosi, "Dynamically exploiting narrow width operands to improve processor power and performance," in *Proc. 5th Int. Symp. on High-Performance Computer Architecture (HPCA-5)*, Jan. 1999.

[38] D. M. Brooks, V. Tiwari, and M. Martonosi, "Wattch: a framework for architectural-level power analysis and optimizations," in *Proc. 27th Int. Symp. Computer Architecture (ISCA-27)*, 2000.

[39] T. D. Burd and R. W. Brodersen, "Energy efficient CMOS microprocessor design" *HICSS*, 1995.

[40] D. Burger, and T. M. Austin, "The SimpleScalar tool set, version 2.0," *SIGARCH Comput. Arch. News*, Vol. 25, Jun. 1997.

[41] J. A. Butts and G. S. Sohi, "A static power model for architects," in *Proc. 33rd Annual IEEE/ACM Int. Symp. on Microarchitecture (MICRO-33)*, Dec. 2000.

[42] A. Buyuktosunoglu, D. Albonesi, S. Schuster, D. Brooks, P. Bose, and P. Cook, "A circuit level implementation of an adaptive issue queue for power-aware microprocessors," in *Proc. Great Lakes Symp. on VLSI Design*, 2001.

[43] B. Calder, D. Grunwald, and J. Emer, "Predictive sequential associative cache," in *Proc. 2nd Int. Symp. on High-Performance Computer Architecture (HPCA-2)*, pp. 244–254, 1996. doi:10.1109/HPCA.1996.501190

[44] R. Canal, A. González, and J. E. Smith, "Very low power pipelines using significance compression" in *Proceedings of the 33rd International Symp. on Microarchitecture*, pp. 181–190, Monterrey, CA, Dec 2000.

[45] R. Canal and A. González, "A low-complexity issue logic," in *Proc. 2000 Int. Conf. on Supercomputing*, pp. 327–335, May 2000. doi:10.1145/335231.335263

[46] —"Reducing the complexity of the issue logic, in *Proc. 2001 Int Conf. on Supercomputing*, June 2001.

[47] A. P. Chandrakasan, S. Sheng, and R. W. Brodersen, "Low-Power CMOS Digital Design", J. Solid-State Circuits, Vol. 27, No. 4 April 1992, pp. 473–484.

[48] J. H. Chang, H. Chao, and K. So, "Cache design of a sub-micron CMOS System/370," in *14th Annual International Symposium on Computer Architecture, SIGARCH Newsletter*, 208–213, IEEE, June 1987.

[49] B. Chatterjee, B. Chatterjee, M. Sachdev, S. Hsu, R. Krishnamurthy, and S. Borkar, "Effectiveness and scalling trends of leakage control techniques for Sub-130 nm CMOS technologies," in *Proc. Int. Symp. on Low Power Electronics and Design (ISLPED)*, 2003.

[50] R. S. Chau, "Intel's breakthrough in high-k gate dielectric drives Moore's law well into the future," http://www.intel.com/technology/magazine/silicon/it01041.pdf.

[51] D. Chen, E. Peserico, and L. Rudolph, "A dynamically partitionable compressed cache," *Singapore-MIT Alliance Symposium*, 2003.

[52] X. Cheng and M. S. Hsiao, "Region-level approximate computation reuse for power reduction in multimedia applications," in *Proc. Int. Symp. on Low Power Electronics and Design (ISLPED)*, pp. 119–122, 2005.

[53] J. H. Chern, J. Jurang, L. Arledge, P. Li, and P. Yang, "Multilevel metal capacitance

models for CAD design synthesis systems," *IEEE Electron Dev. Lett.*, Vol. 13, pp. 32–34, Jan. 1992. doi:10.1109/55.144942

[54] R. C. Chu, R. E. Simons, and G. M. Chrysler, "Experimental investigation of an enhanced thermosyphon heat loop for cooling of a high performance electronics module," *Proc. 15th Semiconductor Thermal Measurement and Management Symposium (IEEE SEMI-THERM)*, March 1999, pp. 1–9.

[55] D. Citron and L. Rudolph, "Creating a wider bus using caching techniques," in *Proc. of IEEE Symp on High Performance Computer Architecture*, pp. 90–99, Jan. 1995. doi:10.1109/HPCA.1995.386552

[56] D. Citron, D. Feitelson, and L. Rudolph, "Accelerating multi-media processing by implementing memoing in multiplication and division units," in *Proc. 8th Int. Conf. on Architectural Support for Programming Languages and Operating Systems (ASPLOS-8)*, 1998.

[57] J. Clabes, J. Friedrich, M. Sweet, J. DiLullo, S. Chu, D. Plass, J. Dawson, P. Muench, L. Powell, M. Floyd, B. Sinharoy, M. Lee, M. Goulet, J. Wagoner, N. Schwartz, S. Runyon, G. Gorman, P. Restle, R. Kalla, J. McGill, and S. Dodson, "Design and implementation of the POWER5 Microprocessor," in *Proc. Design Automation Conference (DAC)*, 2004.

[58] L. T. Clark, E. J. Hoffman, J. Miller, M. Biyani, L. Liao, S. Strazdus, M. Morrow, K. E. Velarde, and M. A. Yarch, "An embedded 32-b microprocessor core for low-power and high-performance applications," *IEEE J. Solid-State Circuits*, Vol. 36, No. 11, Nov. 2001. doi:10.1109/4.962279

[59] *Computing in the 21st Century: Nanocircuitry, Defect Tolerance and Quantum Logic [and Discussion]*, R. S. Williams, B. Christianson, Th. Beth. *Philosophical Transactions: Mathematical, Physical and Engineering Sciences*, Vol. 356, No. 1743, *Quantum Computation: Theory and Experiment* (Aug. 15, 1998), pp. 1783–1791. doi:10.1098/rsta.1998.0249

[60] D. A. Connors and W. M. Hwu, "Compiler-directed dynamic computation reuse: rationale and initial results," in *Proc. 32nd Annual IEEE/ACM Int. Symp. on Microarchitecture (MICRO-32)*, 1999.

[61] —"Hardware support for dynamic activation of compiler-directed computation reuse," in *Proc. 9th Int. Conf. on Architectural Support for Programming Languages and Operating Systems (ASPLOS-9)*, 2000.

[62] G. Contreras, and M. Martonosi, "Power prediction for intel XScale® processors using performance monitoring unit events," in *Proc. Proc. Int. Symp. on Low Power Electronics and Design (ISLPED)*, 2005.

[63] V. Degalahal, L. Li, N. Vijaykrishnan, M. Kandemir, and M. J. Irwin, "Soft error issues in low-power caches," *IEEE Transactions on Very Large Scale Integration (VLSI) Systems*, Vol. 13, No. 10, pp. 1157–1166, Oct. 2005. doi:10.1109/TVLSI.2005.859474

[64] S. Dharmapurikar, P. Krishnamurthy, and D. Taylor, "Longest prefix matching using Bloom filters," in *SIGCOMM*, 2003.

[65] A. Dhodapkar, C. H. Lim, G. Cai, W. R. Daasch, "TEM2P2EST: a thermal enabled multi-model power/performance ESTimator," *Proc. First Int. Workshop on Power-Aware Computer Systems-Revised Papers*, pp. 112–125, Nov. 12, 2000.

[66] K. Diefendorff, P. K. Dubey, R. Hochsprung, and H. Scale, "AltiVec extension to PowerPC accelerates media processing" *IEEE Micro*, Mar/Apr 2000.

[67] J. Donald and M. Martonosi, "Techniques for multicore thermal management: classification and new exploration," in *Proc. 33rd Int. Symp. on Computer Architecture (ISCA-33)*, pp. 78–88, 2006. doi:10.1109/ISCA.2006.39

[68] S. Dropsho, A. Buyuktosunoglu, R. Balasubramonian, D. H. Albonesi, S. Dwarkadas, G. Semeraro, G. Magklis, M. L. Scottt, "Integrating adaptive on-chip storage structures for reduced dynamic power," in *Proceedings of the International Conference on Parallel Architectures and Compilation Techniques*, 2002.

[69] S. Dropsho, V. Kursun, D. H. Albonesi, S. Dwarkadas, and E. G. Friedman, "Managing static leakage energy in microprocessor functional units," in *Proc. 35th Annual IEEE/ACM Int. Symp. on Microarchitecture (MICRO-35)*, Dec. 2002.

[70] D. Duarte, N. Vijaykrishnan, M. J. Irwin, H.-S. Kim, and G. McFarland, "Impact of scaling on the effectiveness of dynamic power reduction schemes," in *Proceedings of the 20th International Conference on Computer Design (ICCD)*, Freiburg, Germany, pp. 16–18, Sept. 2002.

[71] D. Dunn, "The best and worst cities for data centers," *Information Week*, Oct. 23, 2006 edition.

[72] O. Ergin, D. Balkan, K. Ghose, D. Ponomarev, "Register packing: exploiting narrow-width operands for reducing register file pressure," *Proc. 37th Annual IEEE/ACM Int. Symp. on Microarchitecture (MICRO-37)*, 2004.

[73] D. Ernst, N. S. Kim, S. Das, S. Pant, R. Rao, Toan Pham Ziesler, C. Blaauw, D. Austin, T. Flautner, K. Mudge, T, "Razor: a low-power pipeline based on circuit-level timing speculation," in *Proc. 36th Annual IEEE/ACM Int. Symp. on Microarchitecture (MICRO-36)*, Dec. 2003.

[74] X. Fan, W. Weber, and L. A. Barroso, "Power provisioning for a warehouse-sized computer," in *Proc. 34th Int. Symp. on Computer Architecture (ISCA-34)*, 2007.

[75] M. Farrens and A. Park, "Dynamic base register caching: A technique for reducing address bus width," in *Proc. 18th Int. Symp. on Computer Architetcure (ISCA-18)*, 1991.

[76] B. Fields, R. Bodik, M. D. Hill, "Slack: maximizing performance under technological constraints" *Proc of the 29th Annual International Symp on Computer Architecture*, 2002.

[77] K. Flautner, N. S. Kim, S. Martin, D. Blaauw, and T. Mudge, "Drowsy caches: simple techniques for reducing leakage power," in *Proc. of the 29th Int. Symp. on Computer Architecture (ISCA-29)*, pp. 148–157, 2002. doi:10.1109/ISCA.2002.1003572

[78] K. Flautner, S. Reinhardt, and T. Mudge, "Automatic performance setting for dynamic voltage scaling," *Wireless Networks* Vol. 8, 2002. doi:10.1023/A:1016546330128

[79] M. J. Flynn, "Deep-Submicron Microprocessor Design Issues," *IEEE Micro*, Vol. 19, No. 4, July/Aug. 1999, pp. 11–22. doi:10.1109/40.782563

[80] D. Folegnani and A, González, "Energy-effective issue logic," in *Proc. 28th Int. Symp. on Computer Architecture*, July 2001.

[81] F. Gabbay and A. Mendelson, "Can program profiling support value prediction?," in *30th International Symposium on Microarchitecture*, 1997.

[82] M. Galluzzi, V. Puente, A. Cristal, R. Beivide, J.-A. Gregorio, M. Valero, "A first glance at Kilo-instruction based multiprocessors," Conf, *Computing Frontiers*, pp. 212–221, 2004.

[83] K. Ghose and M. B. Kamble, "Reducing power in superscalar processor caches using subbanking, multiple line buffers, and bit-line segmentation," *Int. Symp. on Low Power Electronics and Design*, 1999.

[84] S. Gochman, R. Ronen, I. Anati, A. Berkovits, T. Kurts, A. Naveh, A. Saeed, Z. Sperber, and R. Valentine, "The Intel Pentium M Processor: Microarchitecture and Performance," *Intel Tech. J.*, May 2003, pp. 21–36.

[85] R. Gonzalez and M. Horowitz, "Energy dissipation in general purpose microprocessors," *IEEE J. Solid-State Circuits*, Vol. 31, No. 9, Sept. 1996, pp. 1277–1284. doi:10.1109/4.535411

[86] A. González, J. Tubella, and C. Molina, "Trace-level reuse," in *Proceedings of the International Conference on Parallel Processing (ICPP)*, 1999.

[87] M. K. Gowan, L. L. Biro, and D. B. Jackson, "Power considerations in the design of the alpha 21264 microprocessor," *Proc. 35th Design Automation Conference (DAC)*, 1998.

[88] D. Grunwald, A. Klauser, S. Manne, and A. Pleszkun, "Confidence estimation for speculation control," in *Proc. 25th Intl. Symp. on Computer Architecture, (ISCA-25)*, pp. 122–131, 1998.

[89] D. Grunwald, P. Levis, K. I. Farkas, C. B. Morrey, III, and M. Neufeld, "Policies for dynamic clock scheduling," in *Proc. Fourth USENIX Symp on Operating Systems Design and Implementation (OSDI 2000)*.

[90] E. Hallnor and S. Reinhardt, "A unified compressed memory hierarchy," in *Proc. 11th Int. Symp. on High Performance Computer Architecture (HPCA-11)*, 2005.

[91] H. F. Hamann, A. Weger, J. Lacey, Z. Hu, P. Bose, E. Cohen, J. Wakil, "Hotspot-limited microprocessors: direct temperature and power distribution

measurements," *IEEE J. Solid-State Circuits*, Vol. 42 No. 1 Jan. 2007, pp. 56–65. doi:10.1109/JSSC.2006.885064

[92]   A. Hamilton, "Brains that click," *Popular Mechanics*, March 1949.

[93]   H. Hanson, M. S. Hrishikesh, V. Agarwal, S. W. Keckler, and D. Burger, "Static energy reduction techniques for microprocessor caches," *IEEE Trans. and Very Large Scale Integration (VLSI) Systems*, Vol. 11, No. 3, pp. 303–313, June 2003. doi:10.1109/TVLSI.2003.812370

[94]   H. Hanson, S. W. Keckler, and D. Burger, "Static energy reduction techniques in microprocessor caches," *Tech Report TR2001-18*, Computer Architecture and Technology Laboratory, Department of Computer Sciences, The University of Texas at Austin, 2001.

[95]   A. Hasegawa, I. Kawasaki, K. Yamada, S. Yoshioka, S. Kawasaki, P. Biswas, "SH3: high code density, low power," *IEEE Micro*, Vol. 15, No. 6, Dec. 1995. doi:10.1109/40.476254

[96]   J. L. Hennessy, and D. A. Patterson, "*Computer Architecture: A Quantitative Approach*." 4th edition. Morgan Kaufmann, 2006.

[97]   S. Heo, K. Barr, and K. Asanović. "Reducing power density through activity migration," in *Proc. Int. Symp. on Low Power Electronics and Design (ISLPED)*, Aug. 2003.

[98]   S. Heo, K. Barr, M. Hampton, and K. Asanović, "Dynamic fine-grain leakage reduction using leakage-biased bitlines," in *Proceedings of the 29th Annual International Symposium on Computer Architecture*, May 2002.

[99]   M. Hill, "A case for direct-mapped cache," *IEEE Computer*, Vol. 21, No. 12, pp. 25–40, Dec. 1988.

[100] S. Hines, J. Green, G. Tyson, and D. Whalley, "Improving program efficiency by packing instructions into registers," *Proc. 32nd Int. Symp. on Computer Architecture (ISCA-32)*, pp. 260–271, 2005. doi:10.1109/ISCA.2005.32

[101] M. Horowitz, T, Indermaur, and R. Gonzalez, "Low-power digital design," *Proc. IEEE Symp. on Low Power Electronics*, San Diego CA, October 1994, pp. 8–11. doi:10.1109/LPE.1994.573184

[102] HotSpot. http://lava.cs.virginia.edu/HotSpot/.

[103] C. Hsu, and U. Kremer, "The design, implementation, and evaluation of a compiler algorithm for CPU energy reduction", in *Proc. of the ACM SIGPLAN 2003 Conf. on Programming Language Design and Implementation (San Diego, CA, USA, June 09–11, 2003)*, PLDI '03, pp. 38–48. doi:10.1145/781131.781137

[104] J. S. Hu, A. Nadgir, N. Vijaykrishnan, M. J. Irwin, M. Kandemir, "Exploiting program hotspots and code sequentiality for instruction cache leakage management," in *Proc. Int. Symp. on Low Power Electronics and Design (ISPLED)*, 2003.

[105] Z. Hu, A. Buyuktosunoglu, V. Srinivasan, V. Zyuban, H. Jacobson, and P. Bose, "Microarchitectural techniques for power gating of execution units," in *Proc. Int. Symp. on Low Power Electronics and Design (ISLPED)*, 2004.

[106] Z. Hu, P. Juang, P. Diodato, S. Kaxiras, K. Skadron, M. Martonosi, D. W. Clark "Managing leakage for transient data: decay and quasi-static memory cells," in *Proc. Int. Symp. on Low Power Electronics and Design*, 2004.

[107] J. Huang and D. J. Lilja, "Exploiting basic block value locality with block reuse," in *Proc. 5th Int. Symp. on High-Performance Computer Architecture (HPCA-5)*, 1999.

[108] M. C. Huang, J. Renau, S.-M. Yoo, and J. Torrellas, "L1 data cache decomposition for energy efficiency," in *Proc. Int. Symp. on Low Power Electronics and Design (ISLPED)*, pp. 10–15, 2001.

[109] T. Inoue, T. Ishihara, and K. Murakami, "Way-predicting set-associative cache for high performance and low energy consumption," in *Proc. of the International Symposium on Low Power Electronics and Design*, pp. 273–275, 1999.

[110] *Intel Corp. Intel®64 and IA-32 Architectures Optimization Reference Manual*, http://www.intel.com/ design/processor/manuals/248966.pdf

[111] *Intel Corp. Intel®SA-1110 Processor Developer's Manual*. http://developer.intel.com/ design/strong/.

[112] *Intel Corp. Intel®XscaleTM Core Developer's Manual*. http://developer.intel.com/ design/intelxscale/.

[113] *International Technology Roadmap for Semiconductors*, 2005 edition, 2006 update. http://www.itrs.net/

[114] C. Isci and M. Martonosi, "Runtime power monitoring in high-end processors: methodology and empirical data," in *Proc. 36th Annual IEEE/ACM Int. Symp. on Microarchitecture (MICRO-36)*, pp. 93–104, 2003.

[115] C. Isci and M. Martonosi, "Identifying program power phase behavior using power vectors", in *Proc of the IEEE International Workshop on Workload Characterization (WWC6)*, 2003.

[116] C. Isci, G. Contreras, and M. Martonosi, "Live, runtime phase monitoring and prediction on real systems with application to dynamic power management," in *Proc. 39th Annual IEEE/ACM Int. Symp. on Microarchitecture (MICRO-39)*, Dec. 2006.

[117] A. Iyer and D. Marculescu, "Power and performance evaluation of globally asynchronouslocally synchronous processors," *Proc. 29th Int. Symp. Computer Architecture (ISCA-29)*, pp. 158–170, 2002. doi:10.1109/ISCA.2002.1003573

[118] M. G. Johnson Kin and W. H. Mangione-Smith, "The filter cache: an energy-efficient memory structure," in *Proc. 30th Int. Symp. on Microarchitecture*, Nov. 1997.

[119] R. Joseph and M. Martonosi, "Run-time power estimation in high performance micro-processors," in *Proc. Int. Symp. on Low Power Electronics and Design (ISLPED)*, 2001.

[120] N. P. Jouppi, P. Boyle, J. Dion, M. J. Doherty, A. Eustace, R. W. Haddad, R. Mayo, S. Menon, L. M. Monier, D. Stark, S. Turrini, J. L. Yang, R. Hamburgen, J. S. Fitch, and R. Kao, "A 300-MHz, 115W, 32b bipolar ECL microprocessor," *IEEE J. Solid-State Circuits*, Nov. 1993, pp. 1152–1166. doi:10.1109/4.245601

[121] N. P. Jouppi, P. Boyle, and J. S. Fitch, "Designing, packaging, and testing a 300-MHz, 115 W ECL microprocessor," *IEEE Micro*. Vol. 14, No. 2 (April 1994) pp. 50–58. doi:10.1109/40.272837

[122] T. Juan, T. Lang, and J. J. Navarro, "The difference-bit cache," in *Proc. 23rd Annual International Symp. on Computer Architecture (ISCA-23)*, pp. 114–120, 1996. doi:10.1145/232973.232986

[123] M. B. Kamble and K. Ghose, "Analytical energy dissipation models for low power caches," in *Proc. Int. Symp. on Low-Power Electronics and Design*, 1997.

[124] J. Kao and A. Chandrakasan, "Dual-threshold voltage techniques for low-power digital circuits," *IEEE J. Solid State Circuits* Vol. 35, 2000. doi:10.1109/4.848210

[125] A. Karandikar and K. K. Parhi, "Low power SRAM design using hierarchical divided bit-line approach," in *Proc. Int. Conf. Computer Design (ICCD)*, Oct. 1998.

[126] T. S. Karkhanis and J. E. Smith, "A first-order superscalar processor model," *Proceedings of the 31st Annual International Symposium on Computer Architecture*, p. 338, June 19–23, 2004. doi:10.1109/ISCA.2004.1310786

[127] S. Kaxiras, Z. Hu, and M. Martonosi, "Cache decay: Exploiting generational behavior to reduce cache leakage power," in *Proc. 28th Int. Symp. on Computer Architecture (ISCA-28)*, 2001.

[128] S. Kaxiras and P. Xekalakis, "4T-decay sensors: a new class of small, fast, robust, and low-power, temperature/leakage sensors," in *Proc. Int. Symp. on Low Power Electronics and Design*, 2004.

[129] S. Kaxiras, P. Xekalakis, and G. Keramidas, "A simple mechanism to adapt leakage-control policies to temperature," in *Proc. Int. Symp. on Low Power Electronics and Design*, 2005.

[130] G. Keramidas, K. Aisopos, and S. Kaxiras, "Dynamic dictionary-based data compression for level-1 caches," *Architecture of Computer Systems*, 2006.

[131] G. Keramidas, P. Xekalakis, and S. Kaxiras, "Applying decay to reduce dynamic power in set-associative caches," in *Proceedings of the High-Performance Embedded Architetcure and Compilation Conference*, 2007.

[132] A. Keshavarzi, S. Ma, S. Nagendra, B. Bloechel, K. Mistry, T. Ghani, S. Borkar, and V. De, "Effectiveness of reverse body bias for leakage control in scaled dual Vt CMOS

ICs," in *Proc. Int. Symp. on Low Power Electronics and Design (ISLPED)*, pp. 207–212, 2001.

[133] R. E. Kessler, R. Jooss, A. Lebeck, and M. Hill, "Inexpensive implementations of set-associativity," in *Proceedings of the 16th Annual International Symposium on Computer Architecture*, pp. 131–139, May 1989. doi:10.1109/ISCA.1989.714547

[134] R. E. Kessler, "The Alpha 21264 microprocessor," *IEEE Micro*, Vol. 19, No. 2, pp. 24–36, Mar/Apr 1999. doi:10.1109/40.755465

[135] M. Kharbutli and Y. Solihin, "Counter-based cache replacement algorithms," in *Proc. 2005 International Conference on Computer Design*, 2005.

[136] N. S. Kim, T. Austin, D. Blaauw, T. Mudge, K. Flautner, J. S. Hu, M. J. Irwin, M. Kandemir, and V. Narayanan, "Leakage current: Moore's law meets static power," *Computer*, Vol. 36, pp. 68–75, 2003.

[137] N. S. Kim, K. Flautner, D. Blaauw, T. Mudge, "Circuit and microarchitectural techniques for reducing cache leakage power," *IEEE Transactions on Very Large Scale Integration (VLSI) Systems*, Feb. 2004

[138] N. S. Kim, K. Flautner, D. Blaauw, and T. Mudge, "Drowsy instruction caches–leakage power reduction using dynamic voltage scaling," in *Proc. 33rd Annual ACM/IEEE International Symposium on Microarchitecture (MICRO-35)*, Nov. 2002.

[139] N. S. Kim, K. Flautner, D. Blaauw, and T. Mudge, "Single-V_{dd} and single-V_T super-drowsy techniques for low-leakage high-performance instruction caches," in *Proc. Int. Symp. on Low Power Electronics and Design (ISLPED)*, pp. 54–57, 2004.

[140] C. Kim, D. Burger, and S. W. Keckler, "An adaptive, non-uniform cache structure for wire-delay dominated on-chip caches," in *Proc. 10th Int. Conf. on Architectural Support for Programming Languages and Operating Systems (ASPLOS-10)*, 2002.

[141] D. Kim, T. Austin, and T. Mudge, "Low-energy data cache using sign compression and cache line bisection," *Workshop on Memory Performance Issues*, 2002.

[142] J. Kin, M. Gupta, and W. H. Mangione-Smith, "The filter cache: an energy efficient memory structure," in *Proc. 30th Annual IEEE/ACM International Symp on Microarchitecture, MICRO-30*, 1997.

[143] P. Ko, J. Huang, Z. Liu, and C. Hu, "BSIM3 for analog and digital circuit simulation," in *Proc. IEEE Symp. on VLSI Technology CAD*, pp. 400–429, Jan. 1993.

[144] J. Kolodzey, CRAY-1 Computer Technology, *IEEE Trans.* Components, Packag., Manuf. Technol., Part A, B, C, Jun 1981, Vol. 4, No. 2. pp. 181–186.

[145] G. Kucuk, K. Ghose, D. V. Ponomarev, and P. M. Kogge, "Energy-efficient instruction dispatch buffer design for superscalar processors," in *Proc. Int. Symp. on Low Power Electronics and Design (ISLPED)*, 2001.

[146] R. Kumar, D. M. Tullsen, N. P. Jouppi, and P. Ranganathan, "Heterogeneous chip multiprocessors", *IEEE Computer*, Vol. 38, No. 11, pp. 32- 38, Nov. 2005.

[147] R. Kumar, K. I. Farkas, N. P. Jouppi, P. Ranganathan, and D. M. Tullsen, "Single-ISA heterogeneous multi-core architectures: The potential for processor power reduction," in *Proc. 36th Annual IEEE/ACM Int. Symp. on Microarchitecture, (MICRO-36)*, Dec. 2003.

[148] S. Y. Larin, "Exploiting program redundancy to improve performance." Ph.D. thesis, N. Carolina State University, 2000.

[149] A. R. Lebeck, J. Koppanalil, T. Li, J. Patwardhan, and E. Rotenberg, "A large, fast instruction window for tolerating cache misses," in *Proc. 29th Int. Symp. on Computer Architecture (ISCA-29)*, 2002.

[150] L. H. Lee, B. Moyer, and J. Arends, "Instruction fetch energy reduction using loop caches for embedded applications with small tight loops," in *Proc. Int. Symp. on Low Power Electronics and Design (ISLPED)*, 1999.

[151] H.-H. S. Lee, G. S. Tyson, and M. K. Farrens, "Eager Writeback – A Technique for Improving Bandwidth Utilization", in *Proc. 33rd Annual IEEE/ACM Int. Symp. on Microarchitecture (MICRO-33)*, Dec. 2000.

[152] H. Li, S. Bhunia, Y. Chen, T. N. Vijaykumar, and K. Roy, "Deterministic clock gating for microprocessor power reduction," in *Proc. 9th Int. Symp. on High-Performance Computer Architecture (HPCA-9)*, 2003.

[153] J. Li and J. F. Martinez, "Dynamic power-performance adaptation of parallel computation on chip multiprocessors," in *Proccedings of the Twelfth International Symp. on High-Performance Computer Architecture*, 2006, Feb. 2006.

[154] L. Li, I. Kadayif, Y. F. Tsai, N. Vijaykrishnan, M. Kandemir, M. J. Irwin, and A. Sivasubramaniam, "Leakage Energy Management in Cache Hierarchies," in *Proc. 2002 Inte. Conf. on Parallel Architectures and Compilation Techniques*, pp. 131–140, 2002. doi:10.1109/PACT.2002.1106012

[155] Y. Li, D. Parikh, Y. Zhang, K. Sankaranarayanan, M. Stan, and K. Skadron, "State-preserving vs. non-state-preserving leakage control in caches," in *Proc. Conf. on Design, Automation and Test in Europe (DATE)*, pp. 22–27, Feb. 2004.

[156] M. H. Lipasti and J. P. Shen, "Exceeding the dataflow limit via value prediction," in *Proc. 29th Annual IEEE/ACM Int. Symp. on Microarchitecture (MICRO-29)*, 1996.

[157] M. H. Lipasti, C. B. Wilkerson, and J. P. Shen, "Value locality and load value prediction," *Architectural Support for Programming Languages and Operating Systems*, 1996.

[158] L. Liu, "Cache designs with partial address matching," in *Proc. 27 Internaltional Symposium on Microarchitecture*, pp. 128–136, 1994.

[159] C.-K. Luk, R. Cohn, R. Muth, R. Muth, H. Patil, A. Kaluser, G. Lowney, S. Wallace, V.J. Reddi, and K. Hazelwood, "PIN: Building customized program analysis tools with dynamic instrumentation", in *Proc. of PLDI 05*, June 2005.

[160] J. Lyman, "Special Report–Supercomputers demand innovation in packaging and cooling," *Electronics*, Sept. 1982, pp. 136–143.

[161] S. Manne, A. Klauser, and D. Grunwald, "Pipeline gating: speculation control for energy reduction," in *Proc. 25th Int. Symp. on Computer Architecture (ISCA-25)*, 1998.

[162] J. Markoff and S. Hansell, "Hiding in plain sight," *Google Seeks More Power*, New York Times, Jun. 14, 2006.

[163] S. M. Martin, K. Flautner, T. Mudge, and D. Blaauw, "Combined dynamic voltage scaling and adaptive body biasing for lower power microprocessors under dynamic workloads," in *Int. Conf. Computer Aided Design (ICCAD)*, 2002.

[164] Y. Meng, T. Sherwood, and R. Kastner, "On the limits of leakage power reduction in caches," in *Proc. 11th Int. Symp. on High-Performance Computer Architecture (HPCA-11)*, pp. 154–165, 2005.

[165] F. J. Mesa-Martinez, J. Nayfach-Battilana, and J. Renau, "Power model validation through thermal measurements," in *Proc. 34th Int. Symp. on Computer Architecture (ISCA-34)*, pp. 302–311, 2007. doi:10.1145/1250662.1250700

[166] P. Michaud and A. Seznec, "Data-flow prescheduling for large instruction windows in out-of-order processors," in *Proc. 7th Int. Symp. on High-Performance Computer Architecture (HPCA-7)*, pp. 27–36, Jan. 2001. doi:10.1109/HPCA.2001.903249

[167] P. Michaud, Y. Sazeides, A. Seznec, T. Constantinous, and D. Fetis, "An analytical model of temperature in microprocessors," *Research Report RR-5744*, INRIA, Nov. 2005.

[168] R. Min, W. B. Jone, Y. Hu, "Location cache: A low-power L2 cache system," in *Proc. of the International Symposium on Low Power Electronics and Design*, 2004.

[169] J. Montanaro, R. T. Witek, K. Anne, A. J. Black, E. M. Cooper, D. W. Dobberpuhl, P. M. Donahue, J. Eno, W. Hoeppner, D. Kruckemyer, T. H. Lee, P. C. M. Lin, L. Madden, D. Murray, M. H. Pearce, S. Santhanam, K. J. Snyder, R. Stehpany, and S. C. Thierauf, "A 160-MHz, 32-b, 0.5 W CMOS RISC microprocessor," *IEEE J. Solid-State Circuits*, Nov. 1996, Vol. 31, No. 11, pp. 1703–1714.

[170] T. Y. Morad, U. C. Weiser, A. Kolodnyt, M. Valero, E. Ayguade. "Performance, power efficiency and scalability of asymmetric cluster chip multiprocessors," *Computer Architecture Letters*, Vol. 5, No. 1, pp. 14–17, Jan.–June 2006.

[171] A. Moshovos, B. Falsafi, and A. Choudhary, "JETTY: filtering snoops for reduced energy consumption in smp servers," in *Proceedings of the 7th International Symposium on High-Performance Computer Architecture*, Jan. 2001.

[172] N. Muralimanohar, R. Balasubramonian, and N. Jouppi, "Optimizing NUCA organizations and wiring alternatives for large caches with CACTI 6.0," in *Proc. 40th Annual IEEE/ACM Int. Symp. on Microarchitecture (MICRO-40)*, Chicago, Dec. 2007.

[173] E. Musoll, T. Lang, and J. Cortadella, "Working zone encoding for reducing the energy in microprocessor address buses," *IEEE Trans. VLSI Systems*, 6(4):568–572, 1998.

[174] S. Mutah, T. Douseki, Y. Matsuya, T. Aoki, S. Shigematsu, and J. Yamada, "1-V Power Supply High-Speed Digital Circuit Technology with Multi-Threshold Voltage CMOS," *IEEE J. Solid-State Circuits*, Vol. 30, No. 8, pp. pp. 847–853, 1995. doi:10.1109/4.400426

[175] *Hiding in Plain Sight, Google Seeks More Power*, J. Markoff and S. Hansell, *New York Times*, June 14, 2006.

[176] R. M. Owens, H. Mehta, and M. J. Irwin, "Some issues in gray code addressing," in *Proc. Great Lakes Symp. VLSI*, Mar. 1996, pp. 178–180.

[177] S. Palacharla, N. P. Jouppi, and J. E. Smith, "Complexity-effective superscalar processors," in *Proc. 24th Int. Symp. on Computer Architecture*, pp. 206–218, June 1997.

[178] D. Parikh, Y. Zhang, K. Sankaranarayanan, K. Skadron, and M. Stan, "Comparison of state-preserving vs. non-state-preserving leakage control in caches," *Workshop on Duplicating, Deconstructing and Debunking (held in conjunction with ISCA-30)*, pp. 14–25, 2003.

[179] I. Park, C. L. Ooi, and T. N. Vijaykumar, "Reducing design complexity of the load/store queue," in *Proc. 36th Annual IEEE/ACM Int. Symp. on Microarchitecture (MICRO-36)*, 2003.

[180] J-K. Peir, S. Lai, S. LU, J. Stark, and K. Lai, "Bloom filtering cache misses for accurate data speculation and prefetching," in *Proc. Int. Conference Supercomputing*, June 2002.

[181] S. Petit, J. Sahuquillo, J. M. Such, and D. Kaeli, "Exploiting temporal locality in drowsy cache policies," in *Conf. on Computing Frontiers (CF'05)*, Ischia, Italy, pp. 371–377, 2005. doi:10.1145/1062261.1062321

[182] D. Ponomarev, G. Kucuk, and K. Ghose, "Reducing power requirements of instruction scheduling through dynamic allocation of multiple datapath resources," in *Proc. of the International Symp. on Microarchitecture*, Dec. 2001.

[183] M. Powell, A. Agrawal, T. Vijaykumar, B. Falsafi, and K. Roy, "Reducing set-associative cache energy via way-prediction and selective direct-mapping," in *Proc. 34th Annual IEEE/ACM Int. Symp. on Microarchitecture (MICRO-34)*, pp. 54–65, Dec. 2001.

[184] M. Powell, S.-H. Yang, B. Falsafi, K. Roy, and T. N. Vijaykumar, "Gated-V_{dd}: a circuit technique to reduce leakage in deep-submicron cache memories," in *International Symposium on Low Power Electronics and Design (ISPLED'00)*, Rapallo, Italy, pp. 90–95, 2000. doi:10.1145/344166.344526

[185]   P. Pujara and A. Aggarwal, "Restrictive compression techniques to increase level 1 cache capacity," *International Conference on Computer Design*, 2005.

[186]   S. E. Raasch, N. L. Binkert, and S. K. Reinhardt, "A scalable instruction queue design using dependence chains," in *Proc of the 29th Annual International Symp. on Computer Architecture*, 2002.

[187]   S. K. Raman, V. Pentkovski, and J. Keshava, "Implementing streaming SIMD extensions on the Pentium III processor," *IEEE Micro*, Jul/Aug 2000.

[188]   S. Ramprasad, N. R. Shanbhang, and I. N. Hajj, "A coding framework for low-power address and data busses," *IEEE Trans. VLSI Systems*, Vol. 7, No. 2, June, 1991.

[189]   P. Ranganathan, S. Adve, and N. P. Jouppi, "Reconfigurable caches and their application to media processing," in *Proceedings of the 27th International Symposium on Computer Architecture*, 2000.

[190]   R. M. Rao, J. L. Burns, A. Devgan, and R. B. Brown, "Efficient techniques for gate leakage estimation," in *Proc. Int. Symp. on Low Power Electronics and Design (ISLPED)*, 2003.

[191]   G. Reinman and N. P. Jouppi, "CACTI 2.0: an integrated cache timing and power model," *Compaq Technical Report #2000/7*, Feb. 2000, Compaq Western Research Lab.

[192]   O. Rochecouste, G. Pokam, and A. Seznec, "A case for a complexity-effective, width-partitioned microarchitecture," *ACM Trans. Architecture Code Optim.* Vol. 3 , No. 3, Sep. 2006.

[193]   E. Rotenberg, J. Smith, and S. Bennett, "Trace cache: A low latency approach to high bandwidth instruction fetching," in *Proc. 29th Annual IEEE/ACM Int. Symp. on Microarchitecture (MICRO-29)*, 1996.

[194]   M.-N. Sabry, "High-precision compact-thermal models," *IEEE Trans. Components Packag. Technol.*, Vol. 28, No. 4, pp. 623–629, Dec. 2005. doi:10.1109/TCAPT.2005.859666

[195]   T. Sakurai and A. R. Newton, "Alpha-power law MOSFET model and its applications to CMOS inverterdelay and other formulas," *IEEE J. Solid-State Circuits*, Vol. 25, No. 2, pp. 584–594, Apr. 1990. doi:10.1109/4.52187

[196]   H. Sanchez, B. Kuttanna, T. Olson, M. Alexander, G. Gerosa, R. Philip, and J. Alvarez, "Thermal management system for high performance PowerPC™microprocessors," in *Proc. 42nd IEEE Int. Computer Conf. (Feb. 23–26, 1997)*, COMPCON, IEEE Computer Society, Washington, DC, p. 325.

[197]   H. Saputra, M. Kandemir, N. Vijaykrishnan, M. Irwin, J. Hu, C.-H. Hsu, and U. Kremer. "Energy-conscious compilation based on voltage scaling", in *Joint Conference on Languages, Compilers, and Tools for Embedded Systems (LCTES'02) and Software and Compilers for Embedded Systems (SCOPES'02)*.

[198] Y. Sazeides, R. Kumar, D. M. Tullsen, and T. Constantinou, "The danger of interval-based power efficiency metrics: When worst is best," *IEEE Computer Architecture Letters*, Vol. 4, No. 1, 2005. doi:10.1109/L-CA.2005.2

[199] G. Semeraro, D. H. Albonesi, S. G. Dropsho, G. Magklis, S. Dwarkadas, and M. L. Scott, "Dynamic frequency and voltage control for a multiple clock domain microarchitecture", *Proc. of the 35th Annual ACM/IEEE International Symp. on Microarchitecture*, 2002.

[200] G. Semeraro, G. Magklis, R. Balasubramonian, D. H. Albonesi, S. Dwarkadas, and M. L. Scott, "Energy-efficient processor design using multiple clock domains with dynamic voltage and frequency scaling," *High-Performance Computer Architecture*, 2002.

[201] J. S. Seng, E. S. Tune, and D. M. Tullsen, "Reducing power with dynamic critical path information," *Micro 34*, 2001.

[202] S. Sethumadhavan, R. Desikan, D. Burger, C. R. Moore, and S. W. Keckler, "Scalable hardware memory disambiguation for high ILP processors," in *Proc. 36th Annual IEEE/ACM Int. Symp. on Microarchitecture (MICRO)*, Dec. 2003.

[203] K. Shevory, "Cultivating server farms," Oct. 25, 2006. *New York Times newspaper*.

[204] P. Shivakumar and N. P. Jouppi. "CACTI 3.0: An integrated cache timing, power, and area model," *WRL Research Report*, 2001.

[205] N. Sirisantana, L. Wei, and K. Roy, "High-perfomance low-power CMOS circuits using multiple channel length and multiple oxide thickness," in *Proc. ICCD*, 2000.

[206] K. Skadron, M. R. Stan, W. Huang, S. Velusamy, K. Sankaranarayanan, and D. Tarjan. "Temperature-Aware Microarchitecture," in *Proc. 30th Int. Symp. on Computer Architecture (ISCA-30)*, pp. 2–13, June 2003.

[207] K. So and R. N. Rechtschaffen, "Cache operations by MRU change," *IEEE Trans. on Comput.*, Vol. 37, No. 6, pp. 700–709, June 1988. doi:10.1109/12.2208

[208] A. Sodani and G. S. Sohi, "Dynamic instruction reuse," in *Proc. 24th Int. Symp. on Computer Architetcure (ISCA-24)*, 1997.

[209] G. S. Sohi, S. Breach, and T. N. Vijaykumar, "Multiscalar processors," in *Proc. 22nd Int. Symp. on Computer Architecture (ISCA-22)*, 1995.

[210] B. Solomon, A. Mendelson, D. Orenstien, Y. Almog, and R. Ronen, "Micro-operation cache: a power aware frontend for variable instruction length ISA," in *Proc. Int. Symp. on Low Power Electronics and Design (ISLPED)*, 2001.

[211] V. Srinivasan, D. Brooks, M. Gschwind, P. Bose, V. Zyuban, P. N Strenski, and P. G. Emma, "Optimizing pipelines for power and performance," in *Proc. 35th Annual IEEE/ACM Int. Symp. on Microarchitecture (MICRO-35)*, Nov. 2002.

[212] M. R. Stan and W. P. Burleson, "Bus-invert coding for low power I/O," *IEEE Trans. VLSI Systems*, Vol. 3, No. 1, March 1995. doi:10.1109/92.365453

[213] C. Su and A. Despain, "Cache designs for energy efficiency," in *Proc. 28th Hawaii Int. Conf. on Systems Science*, 1995.

[214] D. C. Suresh, B. Agrawal, J. Yang, and W. Najjar, "Tunable bus encoder for off-chip data buses," in *Proc. Int. Symp. on Low Power Electronics and Design (ISLPED)*, 2001.

[215] D. C. Suresh, B. Agrawal, J. Yang, W. Najjar, and L. Bhuyan, "Power efficient encoding techniques for off-chip data buses," *International Conference on Compilers, Architecture, and Synthesis of Embedded Systems (CASES)*, 2003.

[216] E. Talpes and D. Marculescu, "Toward a multiple clock/voltage island design style for power-aware processors," *IEEE Trans. VLSI Syst* Vol 13, No. 5, pp. 591–603, 2005. doi:10.1109/TVLSI.2005.844305

[217] D. Tarjan, S. Thoziyoor, and N. P. Jouppi, "CACTI 4.0," *Hewlett-Packard Laboratories Technical Report #HPL-2006-86*, June 2006.

[218] V. Tiwari, S. Malik, and P. Ashar, "Guarded evaluation: Pushing power management to logic-synthesis/design computer-aided design of integrated circuits and systems," *IEEE Trans*, Vol. 17, No. 10, pp. 1051–1060, Oct. 1998.

[219] S. Velusamy, K. Sankaranarayanan, D. Parikh, T. Abdelzaher, and K. Skadron, "Adaptive cache decay using formal feedback control," in *Proc. Workshop on Memory Performance Issues, 2002, in Conjunction with ISCA-29*, 2002.

[220] N. Vijaykrishnan, M. Kandemir, M. J. Irwin, H. S. Kim, and W. Ye, "Energy-driven integrated hardware-software optimizations using simplepower," in *27th Annual International Symposium on Computer Architecture*, June 2000.

[221] L. Villa, M. Zhang, and K. Asanović, "Dynamic zero compression for cache energy reduction," in *33rd International Symposium on Microarchitecture, MICRO-33*, 2000. cache energy reduction," in *Proc. 33rd Annual IEEE/ACM Int. Symp. on Microarchitecture (MICRO-33)*, 2000.

[222] L. Wei, Z. Chen, M. Johnson, and K. Roy, "Design and optimization of low voltage high performance dual threshold CMOS circuits," in *Proc. 35th Annual Conference on Design Automation (DAC)*, 1998.

[223] M. Weiser, B. Welch, A. J. Demers, and S. Shenker, "Scheduling for reduced CPU energy," in *Operating Systems Design and Implementation*, pp. 13–23, 1994.

[224] S. J. E. Wilton and N. P. C. Jouppi, "An enhanced cache access and cycle time model," *IEEE J. Solid-State Circuits*, 1996.

[225] D. A. Wood, M. D. Hill, and R. E. Kessler, "A model for estimating trace-sample miss ratios," in *Proc. ACM SIGMETRICS Conference on Measurement & Modeling Computer Systems*, pp. 79–89, San Diego, CA, May 1991. doi:10.1145/107971.107981

[226] Q. Wu, V. J. Reddi, Y. Wu, J. Lee, D. Connors, D. Brooks, M. Martonosi, and D. W. Clark, "A dynamic compilation framework for controlling microprocessor energy

and performance", in *Proc. 38th Annual IEEE/ACM Int. Symp. on Microarchitecture (MICRO-38)*, pp. 271–282, 2005.

[227] Q. Wu, M. Martonosi, D. W. Clark, V. J. Reddi, D. Connors, Y. Wu, J. Lee, and D. Brooks, "Dynamic compiler driven control for microprocessor energy and performance", in *IEEE Micro Special Issue: Top Picks from Computer Architecture Conferences*, Vol. 26, No. 1, Feb. 2006.

[228] Q. Wu, P. Juang, M. Martonosi, and D. W. Clark, "Formal online methods for voltage/frequency control in multiple clock domain microprocessors," in *Proc. 11th Int. Conf. on Architectural Support for Programming Languages and Operating Systems (ASPLOS-11)*, pp. 248–259, 2004. doi:10.1145/1024393.1024423

[229] F. Xie, M. Martonosi, and S. Malik, "Compile-time dynamic voltage scaling settings: opportunities and limits," in P*roc of the ACM SIGPLAN 2003 Conference on Programming Language Design and Implementation* (San Diego, CA, USA, June 09–11, 2003), PLDI '03, pp. 49–62.

[230] F. Xie, M. Martonosi, and S. Malik, "Intraprogram dynamic voltage scaling: bounding opportunities with analytic modeling," *ACM Trans. Archit. Code Optim.* Vol 1, No. 3 (Sep. 2004), pp. 323–367.

[231] L. Yan, J. Luo, and N. K. Jha, "Joint dynamic voltage scaling and adaptive body biasing for heterogeneous distributed real-time embedded systems," *IEEE Trans. Computer-Aided Des. Integrated Circuits Systems*, Vol. 24, No. 7, July 2005.

[232] C. Yang and C. H. Lee, "HotSpot cache: joint temporal and spatial locality exploitation for I-cache energy reduction," in *Proc. Int. Symp. on Low Power Electronics and Design (ISLPED)*, pp. 114–119, 2004.

[233] J. Yang and R. Gupta, "FV encoding for low-power data I/O", in *Proc. of ACM/IEEE International Symp on Low Power Electronics and Design*, Huntington Beach, CA, Aug. 2001.

[234] J. Yang and R. Gupta, "Frequent value locality and its applications," *ACM Trans. Embedded Computing Systems*, 2002.

[235] J. Yang and R. Gupta, "Energy efficient frequent value data cache design," in *Proc. 35th Annual IEEE/ACM Int. Symp. on Microarchitecture (MICRO-35)*, 2002.

[236] J. Yang and R. Gupta, "Frequent value encoding for low power buses," *ACM Trans. Embedded Computing Systems*, 2004.

[237] J. Yang, Y. Zhang, and R. Gupta, "Frequent value compression in data caches," in *Proc. 33rd IEEE/ACM International Symposium on Microarchitecture*, Monterey, CA, December 2000.

[238] S.-H. Yang and B. Falsafi, "Near-optimal precharging in high-performance nanoscale CMOS caches, in *Proceedings of the 36th Annual IEEE/ACM International Symposium on Microarchitecture*, 2003.

[239] S.-H. Yang, M. D. Powell, B. Falsafi, K. Roy, and T. N. Vijaykumar, "An integrated circuit/architecture approach to reducing leakage in deep-submicron high-performance I-caches," in *Proc. Symposium on High Performance Computer Architecture*, 2001.

[240] Y. Ye, S. Borkar, and V. De, "A new technique for standby leakage reduction in high performance circuits," in IEEE symposium on VLSI circuits, pp. 40–41, 1998.

[241] C. Zhang, F. Vahin, J. Yangm, and W. Najjar, "A way-halting cache for low-energy high-performance systems," in *Proc. of the International Symposium on Low Power Electronics and Design(ISLPED)*, 2004.

[242] C. Zhang, X. Zhang, and Y. Yan, "Two fast and high-associativity cache schemes." *IEEE Micro*, Vol. 17, No. 5, pp. 40-49, 1997. doi:10.1109/40.621212

[243] M. Zhang and K. Asanović, "Fine-grain cam-tag cache resizing using miss tags," in *Proc. International Symp. on Low Power Electronics and Design*, 2002.

[244] M. Zhang and K. Asanović, "Highly-associative caches for low-power processors," *Kool Chips Workshop, 33rd International Symp. on Microarchitecture*, Dec. 2000.

[245] W. Zhang, S. Gurumurthi, M. Kandemir, and A. Sivasubramaniam, "ICR: in-cache replication for enhancing data cache reliability," in *Proc. Int. Conf. on Dependable Systems and Networks*, 2003.

[246] W. Zhang, J. S. Hu, V. Degalahal, M. Kandemir, N. Vijaykrishnan, and M. J. Irwin, "Compiler-directed instruction cache leakage optimization," in *Proc. 35th Annual IEEE/ACM Int. Symp. on Microarchitecture (MICRO-35)*, 2002.

[247] Y. Zhang, D. Parikh, K. Sankaranarayanan, K. Skadron, and M. Stan, "HotLeakage: an architectural, temperature-aware model of subthreshold and gate leakage," *Tech. Report CS-2003-05, CS Dept., University of Virginia*, Mar. 2003.

[248] Y. Zhang, J. Yang, and R. Gupta , "Frequent value locality and value-centric data cache design," *Architectural Support for Programming Languages and Operating Systems*, 2000.

[249] Z. Zhu and X. Zhang, "Access-mode predictions for low-power cache design," *IEEE Micro*, Vol. 22, No. 2, pp. 58-71, 2002.

[250] H. Zhou, M. C. Toburen, E. Rotenberg, and T. M. Conte, "Adaptive mode control: a static-power-efficient cache design," in *Proc. PACT 2001*, Sept. 2001.

[251] V. Zyuban, D. Brooks, V. Srinivasan, M. Gschwind, P. Bose, P. N. Strenski, and P. G. Emma, "Integrated analysis of power and performance of pipelined microprocessors," *IEEE Trans. Comput.*, Vol. 53, No. 8, Aug. 2004. doi:10.1109/TC.2004.46

[252] V. Zyuban and P. Kogge, "Optimization of high-performance superscalar architectures for energy efficiency," in *Proc. Int. Symp. on Low Power Electronics and Design (ISLPED)*, 2000.